"Life is a natural consequence…"
Systemic Evolutions

Originally published in 2020 as two books under a pseudonym: *Involved Interpretation* and *Spectrum of Depthless Enthusiasm*. And later, the two books combined under the title, *The Disciplines of Time* (2022). Written by Chris Handrahan 2018-2020.
This Edition:
Systemic Evolutions
By Chris Handrahan
First Published February 2023
Publisher: Christopher Handrahan
Cover Design: Chris Handrahan
Writing and Editing: Chris Handrahan

ISBN-13: 978-1738879229
ISBN Registered under Publisher/Author Name and Imprint with Library and Archives Canada by the author, Christopher Handrahan.

1. Impending Change for Betterment 1

2. Involved Interpretation 23

3. This Larger, Mostly Unobserved World 49

4. Past Curled Inside the Present 67

5. Stratagem Disguised as Rule and Experience............. 85

6. Consequentially Derived Choices 105

7. The Encouraging Deniability of Belief 117

8. Humanity was Once Nonhumanity 129

9. Survival of the Fittest? 147

10. Spectrum of Depthless Enthusiasm 155

    Envisioning a Normal World................................. 155

    Envisioning a Surreal World.................................. 161

    Aspergic Tendencies in The Surreal World 169

    The Spectrum of Personality 204

    Normalcy in This World.. 209

Systems in Natural and Human Realities

SYSTEMIC EVOLUTIONS
CREATIVE NONFICTION

Dedication
for the beginning

1. Impending Change for Betterment

For evolution to properly function, the environment must transform to permit fresh, innovative possibilities to emerge. Eventually, the intersecting constraints of reality must coincide within multiple, competitive localities: furthering a diversity within life for newer solutions to share, re-shape and take root. Changes in the environment are often results of the product of an environment's evolution: the varied, struggling life that emerges and its by-products. A fully stable environment would limit or exclude evolutionary solutions: as if, through an improbable feat, life could emerge disconnected or cut off from its world so as not to alter it by its actions; and by this consequence to evolve no further as an aspect of life.

With the fundamental, vastly multidimensional connections of forces, organisms, and niches necessary for convergent evolution to take hold, at a point and beyond, it has all repeated through the vigor of development embedded within the world: originality resulting solely through the different manifested expressions of the same essential solutions.

Diversity becomes the expression of evolution rather than solely its solutions, so that diversity of evolutionary traits becomes the constrictions of selected niche-fulfillment rather than pronounced advancement. Inevitably, environments and their coexisting life find a balancing of traits previously unearthed within the possibilities, while plasticity in expressing these traits becomes the new evolution: at least during millennia between increasingly rare, epic evolutionary solutions or external events.

Evolution begins from a simple dimensional solution and progresses to become woven and embedded throughout the created setting accorded to a mutual entanglement of prior forces constraining the successes of developmental solutions for each arriving and surviving species. Life clings not only to a planet but to an entanglement of universal forces for its sustained manifestation and sustenance. Evolution must become prodded into action by external measures revealed or presupposed within the world through the varied representations of its species. It is cumulative of choices, not randomly created but naturally evolved creation based on preconditions of the world it emerges into.

It is the method of the universe: carving away sequential, cumulative potentials into a functional reality. Evolution has no choice, except the selections imposed by the forces enacting reality. Only within the possibilities of environmental niches do choices emerge and grow within aware, life-infused minds and bodies.

A wasteland of evolutionary traits fades and lies buried as a continuous physical radiance of the past and in the present genetic reflections fixed within reality. Traits return as reinvented vestiges, while other times in useful or refined ways and perceptive combinations due to the essential, quiet rapport between a continually fluctuating environment reflective of its clinging life.

–

There can be no complete separation between organisms or species and environments, for otherwise entities would require success to sufficiently advance through randomly charged change completely detached from its consequences. The living organism must reckon with the world constructed by forces and envisioned by balance.

Environments establish diversity, robustness, and predation, inclusive of the localities of species and individuals, in the manufacture of evolution.

Evolution is an aspect, a field, and a force; its essence is the traits or quality of the environment. Without evolution, an environment stifles no less than evolution stifles with limited environments. In this sense, evolution is the environment; and the environment is evolution.

With a requirement of organisms being persuasively attuned with the environment for evolution to occur, so too does the natural world become regulated by its placement within reality, the energy-connected physical world coupled with the universe of light and internal forces, with molecules and compounds constructed of lesser particle units and the inner organization of being forming its peculiar fruition; for it is obvious the universe has evolved for long to attain such immeasurable complexity and stretched order that is *thought* – an evolutionary solution attached to the environment – not to be able to happen by chance alone, and in some arguments needing an infinite of chance to assure this symmetry to occur just once or a small, finite number of times.

Under the specific forces, chance never has much to figure in once evolution truly emerges within a natural, physical environment capable of its growth, for the prior forces of the universe, forming the surfaces upon which the living world acts, constrain its solutions within its sequential breaking of potentials, making a reality which must apparently at least mostly abide by the past, though chance can emerge as an event and manipulative ploy of that reality and its entwined forces.

Once randomness is advanced sufficiently within evolution and its genetic construct within species and the environments, it ceases to be as prominent in selective choices. Potentiality, at this point and further, becomes constrained by an ever more finite quality of available solutions dictated upon newly changing organisms by an established environment and genetic learning: including its predictable ranges of change and degrees of success. These ranges of variations become embedded within the genetic reality of the system as well as the combined environment-organism co-reality through its collective experiences and the retention of information within this cohesive involvement.

The environment is the retention of information afforded for organisms to disclose and select into their predicted future perceptions and solutions, enhancing the robustness of species within the world as they each in conjunction reflect the other's reality.

—

Mind, body, and environment reveal a reciprocated context. For most species harmonized with symbiosis, the environment is a windfall of information, learning, and storage in its mutual relationship. The environment in this way conserves and even eliminates the need for energy requirement by holding it to be found through perceptions; in turn, these perceptions of the environment help shape the physical forms of these species, acting as a means for memory and retention of information through the process of perceiving and acting on those repeating perceptions.

In this way, the connections between mind, bodily shapes and environment creates a unique subjectivity between species and members of species in their afforded perceptions due to niche circumstances.

The environment is an extension of the mind-body due to this utilization: increasing success and conservation of energy. Each species possesses an innately unique sensitivity towards the world. The universe is truly individual in its segregated contexts of sensual, shapely awareness.

Due to the shifting constraints imposed on evolution by the developing environment, continuous traumatic stages overcome and succeed while inevitably leaving everything in a state of recurring and potential transition.

Our fluctuations quickly fade and become absorbed by reality's changing state, yet this state is only one, and ever-present; and evolution is one mere feature or force of this current state of being.

Evolution acts within the bounds constrained upon it by other aspects of the same, shared state, presenting itself as a recent occurrence within reality to adapted entities capable of perceiving it as it willfully and effectually persuades its presence as an exertion upon the world.

Life is an exhibition of energy utilizing the absorption and reflection of light: we, and all species, change due to an evolved stability of animated matter existing within entangled forces. This deceptive solidity is an unparalleled origination of reality: worthy of immense awe and reflection by any contemplative offspring perceptive to their presence.

–

There is accessibility in thought the same as the environment since the two resolve mutually within the world. This perspective-wise obtainability in nature permits thought and thoughtful responses in life, as these essentials are natural and shapely evolved, but not the only expression of a surfacing existence. Such opportunities of potentiality are multifaceted, whether ignored or overlooked, chosen as selected, or reinterpreted. The plasticity of nature and individual species permits selection of its choices to occur in balance with other external selections it encounters within its locality, and each locality intertwined by its broader localities.

The environment, through the continuance of complex life, retains its past impressions of reality condensed within the singularity of a widely skilled present state. The environment is an atrocious exhibition atop a planet in superfluous profundity and sumptuousness, desirous of keen-sighted observers to inevitably seek and find genderless in the finer details.

To accept evolution is to accept the methods of nature that have evolved higher, entangled information-sharing encompassing thought-like complexities and feeling-like networks of instinct. Such thinking and feeling, to achieve such contained densities of emerging realities and transforming into a successful, brimming environment, was an essential benefit that by necessity grew within the swelling opportunities of materiality atop a planet.

In nature as in reality, it is obvious information and instructions pass through and amid the environment. The evolution of organisms, as a process, learns and nurtures through the knowledge of its budding world. There is predetermined exchange through sensory consumption and emotive physicality. It is the same, part of the same process. Evolution can easily adapt this way, rather than everything being isolated: as if separated inside the living being, part of the past, or randomness alone. It is mutually beneficial and unbelievable otherwise from a perspective contained within its occurrence, but it must take time to accumulate within the attentive awareness of species.

This time occurs due to connectedness with other accumulated transcending paths occurring in tandem and along distinctly chosen pathways and scales within overlapping vicinities: beyond and connectable with the originating entity and all other originating entities as a subjective experience of change.

The environment exists through the sensory interconnections of its living beings, as if both particle and wave. These overlapping localities can also interpret as units or localities of time due to the countless pronouncements of selected experience causing change among differing perspectives and perceptions of being, blending it into a wonderful, spirited coincidence capable of survival and playful inspiration.

–

Plants and animals are conceivable, naturally occurring microbial methods of gathering and feasting. On their own, environments can be limited and inhospitable for microbes struggling to benefit from, while resisting, an entropic chaos, and travel is arduous at such scale; but with the opportunity to symbiotically energize animate objects (from the microbial perspective) that will explore and mine the environment for the mutually driven benefit of energy consumption and higher successes.

The planet's surface is a barren landscape of scattered elements and limited, disjointed energy structures. Life, once emerged by the planet, gradually organizes those random elements into new patterns of assembled energy. Energy flow within the environment improves for the forms of life to prosper, and life in turn spreads, furthering the successes of this flow.

The planet is a mere backdrop and the venting innards of embryonic energies. Life creates its own environment: life becomes in conjunction a network of flowing animation for previously unsorted essentials upon planetary surfaces. On a planet, the evolution of life in its entirety is the evolution of a living environment of structure and chemistry atop an evolving surface. A changing, energized surface is essential to advancing evolution, whether microbe, animal, plant, or symbiosis: in physical reality, in the make-up of life and environment, everything is surface atop surface since the material world is surface based at least in the confinement of objects due to the suppressions required to produce vigor or restraint as necessitated.

–

The chance for recurrence in evolution is not necessarily exhausted because it shows no proof during a certain, present *now*, for an evolutionary infancy of the minded seeker of proof may have yet to produce abundant limits to find it.

Life is a natural consequence, dually inhibited by reality's forces and freely willed. It rides a temporal landscape of chemistry, becoming a constructed function compelled in development by environmental forces. Life must tap into the quantum world for energy and conception.

Evolution must restrain and limit selections according to existing forces and their naturally balanced consequences as reflected through the environmental outcome. Evolution does not have a choice except the choices it inevitably makes to succeed.

For more complex organisms to form, the microbial world must first sufficiently evolve in its complexity, no different than the particles and forceful methods of the universe beforehand; as it is with larger complex organisms, they must be given abundant opportunity to exhaust the excessive limits of reality until matured enough through mutual, seemingly endless, error-ridden repetition to overcome and raise itself into an advanced evolutionary spring of better solutions; similar to thought and its inevitable exploration of the limits and extremes of probabilities reaped by a native consciousness attached to the environment it abetted sowing.

–

The planet and its subsequent, evolutionary force of life go through innumerable subtle and sporadically abrupt changes. More generally, these changes flow together gradually permitting adaptation through generations of variation. If ice ages occur over repeating cycles according to the gradual movement of well-adjusted forces, life has adapted through its species having varied guises of development and adaptation based on shifting environments. When the rare occurrence befalls and what should have occurred over millennia abruptly happens within a much-condensed portion of that time into an extreme event, most species fail to possess the necessary representation within its repertoire of hurried transformation.

These interrupting events, in relation to all planetary events, are inevitably overcome and short-lived, and the processes continue to present themselves and continue adapting from an early displaced environment to an eventual less chaotic and less certain one; and the many new species, not much different than before, again learn to transform, where necessary and capable to adapt, per the cycles of the planet, and attempt to advance better solutions once again.

Life on a planet, on its surface, evolved to overcome fleeting catastrophic events that dissipate and eventually revert due to its unsolicited placement within the conditions and forces systematizing its reality. It is how life and the environment, as a continuance, evolved itself. All species combined by a moment, in any one epoch, during any one of large measures of cyclical change can perform the same function and create the same results. This repeats any enormous number of times until it does not anymore. Elsewhere, events of the universe can be the same, even saturated by it; and as with our evolution here, it continues to ebb and flow with tidal-like change.

–

Necessarily, species exist requiring the potential of the individual to *overcome* or better the species, physiologically or psychologically, for its effective evolvement into future successes; otherwise, the species would require to convergently advance and succeed in temporal agreement.

Where an advanced, thought-filled species fears its true overcoming individuals and cheers its false ones, the stunting of continuing solutions is a probable outcome, even to the point of a complete undoing or fatal repression of the species.

Dually, it can be a species' undoing if these succeeding individuals possess moral ambiguity in their influence on the future due to becoming overwrought by an overtly tempered present state of surrounding experiences and surrendering resolves.

–

The organism at the level of cells: inter-compartmentalized with internal workings keeping each cell alive with energy and performance through proteins, and each segment of each single cell with relative, proportional space equal to that of the whole organism; each made of countless individual points of energy themselves with inner working fundamentals. The fine tuning of this reality at such extreme and finite a unit.

The organism constructed by this constrained inner force in multitude, unleashed through evolutionary movements, drives it to find a fit to the underlying environment it merely helps create to reshape.

Where habitable, microbes are a native property of existence as an evolved method to express their energy. It is the same, where the environment permits, with higher evolutionary paths. Yet any consequential higher evolution will remain bound and indebted to its microbial origin. There is no separation without consequence.

Nature withholds overpopulation. In extreme cases, the microbial balance can shift, altering the dominating species due to an inability to keep up with its own distorted environment.

–

The environment is already there when you are born, and you continue along with it in unison. The environment reverberates with the culmination of our past and present entangled choices, its energy awaiting further release or exchange into reality by observation or willful renewal.

There is plentiful source, baser than its higher animated consciousnesses, for the capture and release of the quantum energy needed for this continual replacement of the environment filtered through evolutionary designs of life.

Evolution can enhance innate solutions or choices, such as fighting or fleeing (limited options in such abrupt danger situations anyway), fueling its selective behavior with chemical enhancement. Thereby, in such situations, evolution does not have to create but merely act upon the constraint imposed by the environment or reality. Life must deal with reality, often through insightful, advantageous methods. Evolution acts upon and innovates limited options available for a vast array of presented events. Complexity arises when presented external events become too focused upon a species; then specific aspects of evolution must adapt by whatever means or become overwhelmed and outdone as a species.

Evolution does not make the choices constrained upon life by reality; it just acts upon the usually limited choices given through the restrictions on events placed as properties of reality. The environment should have consideration as the essence of what continues, along with the species or individuals which are supplementary and the face of that environment in its current state.

In nature, species do not find the allowance for overwhelming success. There are always balancing forces, however prolonged, countering this extreme feat.

–

There can only be free will to the extent of limitations due to a bodily form of species, and its need for a fit environment for its behavioral shape to find success. A species needs an environment, even if embedded within something bigger, it is especially suitable according to its selected niche and flexible design.

The external constraints imposed upon existence accorded to its evolution can often supersede, if destructive enough, all evolutionary solutions, resetting all progress for the millennia to find and repeat itself again through long endeavor of change: of once again containing, through species, successful units of moveable, designed and attuned probability created through the attainment and suppression of a workable culmination of successful choices embedded within reality through environmental experience, behavioral routines and genetic reflection.

All the best ideas, the highest thoughts, are simply a likeness or best explanation of reality per the perspective of a single species and its capabilities of perception within a closed environment, with a connection limited by physicality and a false certainty of fullness resulting from past choices of understanding surviving into present believability.

–

Memories create a sense of identity within members of a species: if required due to an affinity for individuality with the current or past evolution of the species as a means of success. The ability to retain specific, unique memories creates the impression of exclusion. In other species less individually absorbed, collective memories find retention in groups whereby sole individual memories exist only as distinction among the group's collective advantage or lineage as shared symbols cuing behaviors. This sense of identity is a consequence of an entity's confidence in its environment and the utility of storing information within the world and most clearly within its selected niche comprehension.

Memories were an original solution affording life the potential to predict the future with accuracy for its survival-benefit. They are a context, in symbolic, symbiotic form, between organisms and the environment. By retaining symbolic representation of the past and its outcomes, including progression or hinderance in prior behavioral feedback, an accurate prediction of the future refines as guide to interactive, niche, or genetic choices.

–

There is no appearance, except that created within the minds of observers in concurrence with the forces of reality. Appearance is the mere interpretation shaped by the hardships of evolution, the elucidation of energized information, collective acceptance for the continuance of succeeded risks, and the appeasement and fondness afforded to minds as a converging trait to recognize reality for its successful advantage and survival.

Appearance is an instrument of subsistence: a predictive force. To image the environment is obviously to have an advantage, safeguarding life by permitting a projecting context to future dangers and external vulnerabilities through past impressions and reoccurring symbols, a connective force utilized by life to recognize an environment for something of what it might closely resemble to an observer.

Each species and individual views a unique aspect of this ready, conjoined appearance and its best interpretation of what this appearance represents according to qualities extracted from the quanta of wave-like, radiating light willingly presented everywhere. In this way, light is the precursor to memory, symbolic recognition, and most subsequent cognitive traits easily usable by a current state of species. Reality, as understood through the various branches of evolution, should not be possible without the discovery and adaptation of organisms ensuing from the useful malleability of light, and its subsequent, motivating effects upon the interacting flows of energy and evolution atop a planet.

–

Motives, impulses, drives, desires, symmetry, sociality: all and everything else humans are synchronistically derived and constructed through a balanced reality to attain and co-exist within a short range of recognition atop a single planet.

Resistance to existence's struggle with a distended chaos is the balance of life's evolution expressed through the retention and advancement of solutions furthering its progress.

Existence is, as essential, contained the same at all levels: whether particle, animal, or planet. These compressions or densities of existence materialize as the embodiment of a retained collection of bound, coincidental solutions.

Evolution requires variation within the members of a species; without it, no species should survive a changing environment long enough to last. This variation, expressed through traits, appearances, subtle differences and abilities or aptitudes, resistances, cultures, allowing species to adapt as a group at the occasional expense of some of its members, rather than an entire, equally-endowed species to disappear in entirety at the first challenge to adaptability by the environment: perhaps a change resulting from an after-effect of its own actions within its niche and thereby its complete undoing as an attempt at a species caused by a singular lacking adaptation due to a helplessness of variation among its members in the face of original environmental change.

In the end, innumerable species have been a necessity of nature, abrupt stirring releases of the undergrowth, often self-destructive and unwitting, fodder for new cycles and trials of being.

Nature is an utmost manipulator of time, with undying patience for the necessities of renewal. The means to an end of evolution is that the continuing present of the past will fertilize the future. Only later, during more advanced states of an environment and the refinement of the traits of species, does evolution have to start increasingly factoring in far-reaching, plausible, or moral effects upon the behaviors and feelings of its advanced assortment of life.

The environment is not a plentifully inert backdrop devoid of influence. Life's reshaping of the environment happens in accord with the environment's reshaping of life, each occurring with no less prominence than the other.

–

There is an evolutionary advantage in being capable of correctly reading or predicting your environment and its occurrences in relation to yourself. When viewing outward the world does not sense in the present, but slightly behind due to local and nonlocal distances; but we also continually predict forward so that our minds are ahead of ourselves as it views an altering perspective of reality encountered as a close aspect of the past. A successful mind must predict urgency in its natural environment while receiving its sensory impressions slightly behind the present: insomuch as the present conditions to exist within such minds continually chasing and chased by it.

The possibilities should exist in our world for species capable of successfully adapting to reality without the necessity of waiting on the past to catch it or a future it needs to predict, while existing solely in a confident, timeless state alongside the universe.

Evolution, on this planet, is not always a seamless, continuous state, but keeps resetting through various external large-scale events. This has been, as likelihood, the case with only few relative exceptions elsewhere in the universe until such a present occurs that these catastrophic events become rare, whether initiated by the continuing effects of the collapsing matter of the universe or a simple locality within it through adaptation or exclusion.

With evolution, randomness is the utilization of an early, redundant stage, later constructed into more directed genetic choices or regulations.

To act on instinct is to follow the more probable-constructed route of action. Much like with the forces of the universe combined symmetrically to exhibit reality, the forces acting on the mind-body of an organism, plant or animal, are a symmetry of bodily form, traits and needs; as a species and an individual genetic idiosyncrasy, the quality of learning and experience it encounters enabling it to predict future probabilities, and its exposure and adaptability to environmental changes potentially caused to its local niche by its own species and other overlapping niches of reality, all determine the proficiency of its instinctual response to the world.

Instinct is not equal in all members of a species but is there to be harmoniously skilled and nurtured into reality after emerging and blending within its environment.

Instinct resembles the actions of a being which must repeatedly choose from limited choices as it routinely makes its way through the course of a life traversing a preferably stable, shared environment; but these choices can often succeed only so far as instinct is freed, permitting it to absorb through experiential perceptions and naturally act within its bodily shape with skill.

The confidence of entity and environment creates an ability to use multidimensional aspects of the environment combined with bodily capabilities to invent and bring into reality finite solutions through creative behaviors, which, once inevitably confronted with annihilation, fade back into the wave of reality where only potential remains for localized re-emergences to temporally occur.

The reciprocal influences between mind-body and environment, as an attracting dynamism of reality, exist as potential information within a locality no different than physical and universal forces intertwine to produce and utilize, as essence, the equivalent information. It is the same with affordance or obtainability in nature: symbols awaiting recognition through unconscious and eventual conscious contact.

These contacts must caress from a shyer nature, as a sensitivity shared in coexistence between reality and the confident individual existing in full belief of itself and its species.

–

An unfortunate fate for humanity (and all similar efforts else-when, -where): not to evolve smart enough fast enough. Luckily, nature does clean up after itself, never afraid of starting over from scratch, complete with the risk of it all, after countless millennia of repeated trial, being probable of reaching the same fate.

Dominating successes of a single species or aspect of nature can appear as a limitation or failing point in the evolution of the world, and as a measurement the true thriving successes of nature occur during the intervals until a dominating species presents itself within the accumulated opportunities.

In this way, an over-dominant species can be a quality of failure for small or broad localities of reality, creating enough imbalance in need of enthusiastic, releasing retreat.

The frustration and tragedy of being too young in evolution to truly recognize reality, to temporally occur where thoughts must become complexly knotted in contest. Simplicity no longer becomes a viable answer. The particles and cells of the world evolve through experiences exposed within the potentialities of a wrought, enacted, and experienced reality.

The mind, especially if self-enabled or freed in its actions, must be a comparable match for the environment, whether expressed solely or communally, otherwise there is no lasting realization.

–

If you learn and follow the ticking of a clock early and long enough, you can come to believe what it tells you is obviously and naturally real as a force rather than a conception of evolutionary advantage meant to gain survival as evolving minds – with an effect so striking it required convergence among most surviving species after its introduction as solution – explored predictive marks contained within the sensitivities of reality vastly capable of a successful strategy for survival; and further elaboration of evolutionary traits as its generational offspring continued to occur.

Traits, especially emotive successes, are capable not through the simple fact of having a trait physiologically appear but also to have it correctly balanced or regulated for functionality reciprocally within the entity and its environment without undo or lethal side-effect.

Whatever landscape life creates is illusory by nature. A simple difference between animate and inanimate lies in the animate possessing the evolved or forced effects enabling movement and range, while inanimate is the residue of animated effects remaining within the universe since its alleged origin.

Reality, and especially life, reaches outside its immediate moment and space beyond its cumulative solutions; also, the universe extends back to contour collective, symmetrical evolution in the expression of species and environments.

–

When you look out over the world, there can be recognition of beauty, and then it stops as a feeling. There is surface, and behind that, there is only more surface.

When infinity first arrived as an idea in a sentient mind wanting an answer, it would ripple instantly outward until impossibly everywhere as a first of its idea, suddenly measuring its probable reality: the abrupt *instance* of infinity everywhere where it never existed before. No answer clarifies a reply. Of course, it would take infinite time to find out if it is true while awaiting an answer: as if quantum plotting of the past is continual, not wrapped up within the present state but floating in a lingering meander in its wake.

With little direct contact with the universe except for dust and light, and an insignificant share of that, a solar system exists for long durations in isolation, barely reaching beyond the threshold of its surfaces.

In looking at the night sky from atop a planet, all that we see exhibited outward in the vast galaxies of stars is fossilized remains of the present state and an isolated concept of infinity to play with.

The only real contact such a spectator could assume within any extended universe would be the tiniest percentage of light this universe can amass as it enters through the threshold of the observer's eye and energizes its mind with the idea of an endlessness in all directions from anywhere it might position itself including exactly where it stands.

The actual universe of the observer of the night sky hardly extends beyond the surface of its bodily form in any attempt to experience that ideal universe it observes from its curled perspective of space. The spectator is genuinely in contact with the universe around it less than a single water molecule contacts an immeasurably sized ocean around it and the life inhabiting it: unless the molecule exhibits the condition of a shared and entangled reality to extend its reach.

–

As consciousness evolves in all life, aspects may select containment, whereby a balance perfects enough without further elaboration. These life forms resist changes from stresses, adapting and existing through outdated traits of evolution. They remain contented unless forcibly disturbed, usually by something highly detrimental and overdue, sometimes to its peril.

Time ceases to be much of a feature in such a balanced existence neglectfully awaiting an endgame to find it: except in the quality of its absence.

Embedded within life and the environment is the mortification of its evolution. The eons of agony and mistakes it takes for species and environments to finally evolve into something eventually resembling a completed, harmonious product; and even if achieved, to exist with such fragility, fears and fullness of horrors due to the long-held choice of evolutionary paths to use feeling as a trait of life within a shared world.

Affordances stemming from the qualities of the environment create new innovations and solutions for life, and thereby new traits and behaviors, rather than internally shaped as exclusively through chance, reduced probability, or randomized genetic mutations. Such mutations which appear initially random are explorations and solutions to a world which genetically is a singular conception to an external context: a visual, emotive mapping of a quantum environment experienced only though the resulting organism as physically real.

The collective environment and its current moment, encompassing past, present, and expectation, feed thoughts. The mind is empty and nonexistent without an environmental framework to encourage it. Of course, the environment also responds to thoughts of it, since it is the totality of the appearance of the niche networks of its organisms: it is within the silence between the two where reality cuts.

Environmental affordances defy belief among disinterested minds, as if such a thing might have and only could exist in the past, pagan and symbolic, but they are fooled into thinking we have overcome primitiveness when it has simply been obscured by an overabundance of false, disconnected thoughts and ideas; and, owing its inclusion with reality, the resulting, molded genomic hypothesis meant as best guess at the external reality's current state its offspring are meant to find accorded to the successes of the past.

Beliefs shared between actions of both the environment and its offspring. The more traits a species is willing to assume through its consciousness, advancing its potential thoughtfulness, the higher should be the degree of obligation it owes back to the environment inclusive of other species.

Unfortunately, this obligation will not equally extend within the thoughts of all individuals with the same passion due to probable, native inequalities of variation within a species during this stage of its movement from past to present as it inevitably consumes the future within minded individuals predicting its fuller understanding and implications.

–

It is conceivable and hardly unimaginable that a universe should have a conscious species considering itself advanced with a vast contingent of individuals thinking its reality to be merely self-real. It rouses sensitivity for the struggle and range of resistance entangled within evolutionary change, especially the gradual development of conscious change.

Organisms and environments conceal an interwoven state within reality embedded through multigenerational experience and genetic exploration, within the processes meant to alter and enhance – through the exploration of reality's forceful potentials intended to intensify flowing variation by the balancing of past consequences with current effects – a singular, unwitting ambition of attaining the world's survivability.

Evolution has no certain end. There is no perfection that cannot have the potential to become externally disturbed through transition into sudden ruin. Life is an inherent means to attain forces otherwise impossible within a concrete space. Life is an impending change for betterment, an endless trial and experiment, an attempt at self-description, at the pride of heedless survival. Life, minimally embedded within reality, resists entropic chaos and thereby is an animated revolt against the ambitious forces of the universe meant to quiet it.

2. Involved Interpretation

An aware mind floats atop an ocean of evolutionary, instinctive existence persuading it; atop a vastness of quantum probability and coinciding forces permitting its reality.

It is plausible that any newly arisen, evolving consciousness attempting to promote itself in its niche world should confuse the skilled emotive properties within life and environments with an otherworld or afterlife, much like seeing shadows on a cave wall several hundred millennia ago as they begin to appear more than the humbling influences and signs of environment. Spectacular beliefs over the span of generations during the following millennia can branch from such an original, perceptive idea of unknown forces acting within the creation of the world. In this way, a growing consciousness begins to misunderstand its own latent, instinctive unconscious as something external to itself and to the physical existence its heedful mind finds symbolizing and accepting as something differently substantial in the textures of reality.

Within the individualized group dynamic, this growing separation of belief in a newly collective detection of nature instils the enhancement of the resulting disconnect with the originating environmental instinct as a new way of noticing the world. The world might become an aspect of awe through witnessing the experiences of the countless plays of light and object prevalent within its existence; or the routines of reality could become marks of wonder, such as the sun or seasons. Change marks with significance. But in all cases, it is the environment which creates the symbols of perceptive consciousness, and from there a semblance of creativity, imagination and invention increases. All species, and all awareness, are a product and aspect of the environment. Without an involved environment, the tangible and intangible cease to exist except as cold inactivity for innate life to terribly attempt to affix anew.

–

In this world, peering outward into the desirable light, there is the observable universe: far-reaching and limited. Perspective-wise, the observable imagines to be outwardly shrinking, fading away, and expanding until there is only the appearance of an unfounded beginning to any newly emerging consciousness. The closer you get to such a fixed, perceptive beginning, the more difficult it becomes for a newly evolved cognizance to determine itself, or to make up concepts of its place in a world. Imagination becomes unnatural as it closely approaches this endless beginning, to its modest betterment: and answerable as a motive for its consciousness. Evolutionary solutions never begin as they pertain to greater use of insight under such context, as no consciousness can arise to appreciate its place within such a universe, but solely within its local environment. There can be lessened philosophy, art, and a shortness of ideas arise as opportunities when a new consciousness sees nothing but itself within a universe with a darkened, starless night sky, and no probable means to ever find or much think it exists to look for. Devoid of science and evolutionary traits marked by the cosmos, this consciousness finds only emptiness, void reflected in the sky, and a locality endlessly centered upon itself.

The affordances available for an advancing consciousness to arise contingently on the preconditions its environment emits; including any state of the universe appreciably understood to be past, current, future: or beginning or end.

–

Environmental symbols are an essential property of the unconscious and prevalent throughout nature but must have fuller recognition through learning by any succeeding cognizant mind. In an analogous system, anything approaching the connotation of truth only exists within the atmospheres of nature. The conscious mind constructs from these building blocks. Inevitably, it permits choices on this quest to overcome its heedful, niche ambitions.

The best truth aware minds can hope to achieve is to unmask, through cognizant endeavor, the means to mimic nature and its own instinctive facet, recreating this truth in the physical world it amasses as a reality: thereby beginning its careless erosion within minds and bodies as a symbol of an abject nature and the dilution of emotive, sensory instincts into dichotomizing beliefs.

For each symbol grasped from reality and made concrete as a conscious symbol, there is a consequential loss of connection or attachment with the original object of that symbol's meaning within the world. Writing, art, philosophy, science, all act within the symbols of reality, usually to the exclusion of that symbol's preceding dwelling within the environment and the meaning it evokes non-consciously.

In this way, the environment is recognized solely to put its meaning to rock, paper and mind, as if the environment is only excavated to extract its symbols for further manipulation within the limits of an extensive range of finite risk to which it becomes evolutionarily exposed as a new reality for the human mind.

This method behaves like a particle rather than a wave of identification. This new adventurous world based on extracted symbols follows the landscape of nature: a consequence of the universe's prior evolution at reducing or limiting an originally immeasurable state into something fathomable, culminating in at least one collective, local awareness atop a planet expressing its version of universal proficiency.

Within this locality of space, on this planet, it has taken and lasted roughly a few billion spins around our sun to expose and interpret a crude meaning of reality through a force of life, revealed through a flowing present state, the means and methods of which are chosen by its highly alert inhabitants according to their willing, intended associations of worldly symbols with committed, connotated facets of extracted truth.

The millennia taken to shape bodies and minds with the symbols of its environment and existence, into assorted niches systematized as species. The design of the human species transforms equally by this interpretation. In the same way, humankind and all life arise not simply by the imagination, but through distinctively connecting with a backdrop: the environmental symbols from which extract new, vaster ventures and thereby niche solutions.

Reality creates degrees of awareness, no different than body shapes and corresponding traits form and coalesce in close synchronicity across generations of unique skillsets. The environment stimulates these signs and symbols attractive to new traits, with the fresh potential of new, furthering controls over reality.

The shared environment comprises awakening sensitivities and inventiveness within organisms since the organisms are the environment. Each is equally inclusive in acute poise. Consciousness is an evolutionary, unifying contrivance utilized by species to differing degrees of proficiencies, taught by its innate connection to the environment, locally and nonlocally, through overlapping ranges of coexisting niches.

An alleged advancing consciousness can just as easily construe to regressed unconscious instinctiveness for what it emphatically loses through the exploit of equilibrium.

—

26

An evolving bodily consciousness assumes the growing tendencies and capacities of existence within the confines of its locality, bringing the possible consequences of advanced suffering perpetrated upon an isolated reality as endless, closed atrocity. Once set down this path, unless it overcomes itself by some coincidence within existence, little can be expected from such sophisticated wayward intention: whereby all its achievements are corrupted by preceding and continuing intentions inevitably probable to arise, yet highly improbable to remain among the fictions and mere beliefs of reality, and so instead being deplorably enacted within reality as if it's a worldly stage for vicious ploy.

Inclinations of mind and feelings such as hatred and indifference, of living without any measure of understanding unless self-glorified and properly honored, should not be rewarded with best access to the present state of reality in which to fertilely grow through generations of cumulative, willed individuals: grouping a collective assembly of predisposed, biased approaches to current and future success.

In the initial stages of an advancing conscious awareness, the mind begins to find symbolic descriptions within reality. These images are then magically recreated from an originating reality into a new reality, such as taking an image from the environment and recreating it into an object or the reshaping of an image upon the contours of a surface: acts meant to transfer and instill associated emotions and concepts into the marks as the script of idyllic world memories.

The natural environment inspires and symbolizes the thoughts of any evolving consciousness, no differently than an artificially reshaped reality might be a preferred inspiration for thought and cuing symbols among later generations without the evoked, emotive attachment the prior instills.

A world of thought without meaningful range of sensitivity restricts certain potentials and increases others, such as the likelihood of dissolution into the thickness of the planet's crusting prior attempts at solutions over enormous millennia.

Once life finds cause through fitting conditions, it begins to explore bodily shapes through the animated combinations that maintain functional success, and later it explores sizes as it evolved from these shapes into eventual gigantic proportions in its evolving struggle with environmental balance. Once sufficiently established, the environment settles into predicted cycles, and given enough opportunity, consciousness can advance from these beginnings to explore its newfound environment by similar means of symbolic instinct used to bond with its preceding, barren lifeless environment, inevitably transforming it into something this new awareness might eventually find beautiful, malevolent, or inspire awe through its life-fueled image.

—

The environment stimulates and encourages the desires and originations of all organisms toward paths of successful niche fulfillment afforded as insightful perceptions acquired through interactions. Without the environment to utilize for direction, entities would exist in isolation, devoid from connections, like one of varied unreal objects floating in a vacuum without contact, yet somehow impossibly able to continue without any extension of self into a world around them, however voluntarily motivated towards focused, instinctive solutions. A portion of the distribution within any organism is wistfully aware, but most traits have no need for serious conscious attention and so go mostly overlooked unless examined as an effect vacant a cause.

Naturally, species must survive and have survived to evolve for millions of years successfully to exist in the current present, and to accomplish this feat, there must be attuned sensitivity with the environment, however varied an involvement that expression might appear to each perspective of life.

As a species existing now, this poise remains to humanity, but successfully we have consciously disassociated ourselves from it and so mostly fail to recognize its reality, instead appearing only during episodes of excess stress, life changes or extreme exertion of mind and body due to an abruptness of new unpredicted experiences.

Long-held, successful habits of species can be difficult to overcome. Significant efforts have tried to protect and further this mindful and emotive detachment within reality. Much of the horrors perpetrated upon the evolving history of humanity has been due to this misinterpretation within reality of the environment's skill and role as a mentor for all evolution; most especially the manipulation employed to advance this corrupted misunderstanding of existence linking any human placement in it as separate, exclusive or special to the universe.

When it comes to an entity's relationship to its environment, the mind-body acts in conjunction: just as the conscious-unconscious act in unison within the same environment and bound by the same ensnarement of forces.

Environments shape the intensity, abstraction, and suppression of awareness. For consciousness to exist as a state of development within a species, there should, to fully embrace the opportune possibilities, be varying examples and zealous states of consciousness, all not being equal according to experiences, learning pressures, and developmental pronouncement: the entirety of a world's many environments in its diversity of awareness among its many niche-fulfilling species.

All other species are also evolving consciousness and shaping it according to niche idiosyncrasies. There is no monopoly among humankind. There is no absence of spirit among nonhumans. All species affix to the same reality. Some of nonhumanity's natural limitations and advantages seem more suitable to higher consciousness than other more limited and privileged species such as human beings: a different perceptive awareness, with greater restrictions of shape to interact with its environment, unmotivated by technology to cooperate with as possibility other than nature's; instead, endowed with an inhibited, focused consciousness, of a kind mostly emulated by less restrained human beings now as a rare idyllic balance of physical and spiritual, as a relaxed state common for other species autonomous within an affixed niche lifestyle.

Debasement of spirit at the expense of the material can happen and might appear as temporal, seductive successes in conscious readings of other species as an overt benefit.

As a starting point, the inestimable presence of a world is much for any new cognizance to break down and overcome if it is to advance mindful thinking. For such a consciousness to fully succeed, it must outwit the countless potentials it is presented by the originating reality; and the countenance of any world realization has much to do with the opportunities afforded by its bodily formula and the incoming sensory illustrations impressed upon it as gifts from the environment shared in multiplicity with other species.

–

Consciousness is a likeness of the alert instinctiveness of nature, just less silent due to eons of tortuous trial and excessive meditation of balancing solutions. The part of the human psyche we identify as conscious now has a worded voice, often giving it the feeling of being imperious, exclusive, and dominating; but then it would like you to think that since it has the voice and is talented with convincing arguments.

In the same way, human realization with its symbolic, linguistic meanings is a phase of this duality in its relationship as a strong, willful voice for an instinctively unconscious atmosphere from which it leapt, and the symmetry of existence permitting its appearance contrary to any alternatives less brilliant. This association with reality unarguably possessed by humans within its potentials of being is not much removed, if at all, from the same nonhuman element and its communicative relationship with reality as a scaling spectrum of involved interpretation.

The unconscious and conscious domains fervently describe dual visages of the same evolving reality, just one is silent and less obvious in its communications, while also being especially complex in attunement to this reality's environmental scape, celestial forces and the density of perceptive minds and bold bodily skills in measured, undulating command of it. Due to scaling perspective, there is no real depth to the universe. There is no need for reductions in defining the textured multiplicity of a multilayered landscape to a succeeding entity attuned to its reality as niche survival.

A deep, prolonged silence of comingling forces imposes its willful direction upon available local events such as planets and their fertile surfaces permitting adaptation to environments through evolutionary, progressive solutions to manifold arising complex problems. This change owes its furtherance to genetic, experiential marks retained within reality as cause for attentive awareness to any species sufficiently evolved to need to witness its own truer reflection within its strange reality.

In character, this permeating, instinctive silence is very much like us and in nonhumans; since we are its image as much as what we think of as our everyday personalities. It is much less obvious to any perceptive too aware of outward distraction to recognize or not impressed with the will to emotively sight it. In nonhumans, the likelihoods of deception are equally ready, though not as mindfully vulnerable due to the chanced lethality of ensuing experience by stepping outside the bounds of an environmental niche.

Humanity, as with nonhumanity, has been about revelation, a material exhibition of willful awareness perpetrated upon the texture of reality, searching the boundaries of consciousness no different than species fill every niche of this environment by similar strategy. It is, as essential, the same: a narrowing of complex affordances into successfully connected behavioral doers; and reapers of what has occurred or predicted as likely to occur.

–

With sensations being the absorption of energy and force, the transfer of information as a likeness of a surrounding environment, reality works through functioning as a system of shared relatedness while still adopting the exclusive inventive realities of differing depths of perceptive insights among individuals of species.

A spectrum of consciousness distributed as species share reality while retaining distinct ownership. Each species contours and texturizes its private reality according to the contribution of innovative beliefs traced by all other individuals occupying the same enclosed milieu as the current, sensitive culmination of conjoint, popular experiences.

Dreams can be the frantic imagery of the evidence within experiences the owner failed to other than latently recognize during everyday waking reality: sardonic of a mindful failure to notice or understand any meanings, penalties or intentions attached to these experiences while they occurred.

Often, minds heavily laden with daily mindful neglect will fail to believe in coincidences, entanglements, or immediate attunement within environments, and thereby fail to interpret the world around them except through the jolting actions of dreams summoned from ancient, discarded survivability. Rarely and with enough belief, the meaning behind these symbols can be relevant enough to find expression outside the world of dreams to become inserted into reality for fathoming evidence and an only source of proper exhibit accorded the instruction of experience.

–

As conscious awareness accumulates within an upright walking species, it must achieve a position where it looks at its reflected identity as a placed context within history inclusive of the broadened world's present state, wondering on the amass of its ruination of a natural world inclusive of its own species, questioning who it is, what it will and will not be agreed upon as a target within its aiming, flittering capacity for primal, dominating positioning.

Humankind argues to be a mystified, distracted species. At a point in its history it boldly, whether for pride or spitefulness, grabbed an implement from an assortment, a solution other species will tend to shyly ignore, and elude with prey-like precision: but humanity gathered it, greater use of conscious awareness in finding the newfound symbols of survival, whatever it originally thought and still thinks that is, in all its confused, overextended brutality and illusory wonder: the immeasurable experience of taxing all potentials a physically sentient world can express through bodily shapes and embracing environments, each minuscule feature of emotive facility and perverse new spectrums of decadences, acting as an ignorant, comingling fulfillment.

The overwhelming penetration of self-pleasure perpetrated upon a reality simply due to its permission within existence as an indulgent opportunity to manifest; and to suffer the unabated sensitivity to a surfacing nature of environmental detestation as a defining gamut of history and its presently evolving inheritance.

No species should be continually healing and tending to its consciousness through long effort due to uncontrolled, underestimated inclinations against the venerated unconscious reality it inhabits as a facet it can no longer fittingly detect.

Inevitably, to succeed beyond the moment, this consciousness should come to recognize itself the object of reality, part of the natural evolution of an advancing world around it, living within *its* space, founded on *its* forces and sensitivities; to advance a consciousness within an environment is to traverse the labyrinth of an oddly created reality, and the establishment of its new reality can result in an arduously distorted and undesirable exhibition of itself repeating over many millennia of nurturing, recognizable only through hindsight during a much later present state, following countless generations of indulgent reexperiences.

–

The exchange provisioned by affording environments are naturally surfacing symbols, very much an impression upon the imagination, available for association, and encouraging for better designs in role improvement. A treading consciousness requires the use of language and crafted symbols, and the representation of such awareness is not exclusive to human consciousness but widespread within the natural world.

These affording connections within the multilayered participants of niched environments can be non-linguistic assessments, explanations, and communicative reckonings. It sees or finds as a sudden, predictive intuition of an obtainability of reality rather than as descriptive reasoning requiring analytic explanations after an innovative indication or insight, though both are the same duality of awareness.

Confident environments also provide the boundaries of risk as gifts of warning within reality for humans and nonhumans based on the sensitivity of bodily shapes, traits, and a context of instinctive niche attachment to its world.

–

A disharmony which emerges from knowledge and its attained attention is that the emotive, thinking species must live with knowledge too; they consciously strive for greater knowledge, higher imagining, and vaster explorations to find what they attempted with earlier analyses but could not fulfill or maintain. It is cumulative and long-winded, so long as they do not permanently turn away from it and accept more submissive, ready tactics.

This wide-ranging gulf between ignorance and current extents of knowledge provides an attempt to reap unique outcomes within reality through a genuine striving of the conscious species to cleverly find the unconscious nature of its world as the disclosure and reshaping into the idiosyncrasies of a localized, material reality: useful as a shared branching of newly afforded traits appropriated through temporal upright reshaping of a species tended by responsive reverberations feedbacking from the fluctuating environment.

The universe is partially knowable, easy to learn and adapt from a fit internal perspective conferred an evolved species, and methodically difficult as a never-ending labyrinth to fathom in hidden, intricate density. It succeeds in working out complex problems into subsequent solutions, sorted through the vast horizons of findings for the willful purpose of constructing successful networks favoring material opportunities fundamental to the forces of evolutionary life in creating vigorous uniformity as an isolating reality.

Humanity is an unfinished current state of this involvement, evolved to think and become willfully aware, to live out this coexistence and coalesce of actuality as if curled up within space like a diminutive hidden dimension of reality all for itself among the inconceivable immensity of the universe.

The arrival of higher human-minded realization merged with the entire content of the prior world's attained actuality of vigorous, correlated networks. It forced humanity to assess and challenge its boundaries and make them substantial within its newfound, supplanted milieus. The aspects of violence, passions, all calculating actions and intended systems of nature, are slanted by this challenge to disclose the boundaries of a shapely, multifaceted environment embedded in the middle of nowhere surrounded by a seeming eternity, especially and obsessively the impulse to survival by better self-advantage, an imprint of eons of evolution searching for success, variety and improvement to better the probing of bodies and environments for mutual, cohesive harmonization.

There are limits to this illusory side of reality no different than the physical reality itself, and any advancing brainchild of the natural world with a dexterous mind and body shape must inevitably provoke this lasting instinctual antiquity of reality and the societal and ecological impairment it can cause across millennia if left to escalate unchecked into an unsolicited dissolution of attentiveness.

Attentive sensitivity towards reality is not unique to humanity. Looking at nature, with our hypothetical exclusion from it: as an entirety, it operates by similar means as human beings, far exceeding in its scaled measure of accomplishments, capacities and with far greater spans of temporal success.

The environment, whether planet or universe, is the storage and retrieval of information; or, more simply, the environment cues and demands thought by its retrieval, presented there as learning and transformed into vastly diverse activities and inevitable genetic-crafted beliefs of body and mind.

At least, nature is conscious since humanity believes and calls itself conscious, making of us an embedded signature of nature's cognizance; humanity is the present culmination of that devised, precise creation, no different than all other species.

—

There is no point in asking *"why?"* of existence for an answer since the question is only valid to consciousness: the answer does not exist in any other than imaginable context, though the struggle to answer the unanswerable leads to exceptional formulated content which can appear convincingly impassioned. The history of wildly human judgements and sketches of morality are the stages which must be traversed through advancing a consciously held reality, even if mostly it is fraught with errors: though these errors often at least temporarily succeed a functional, lasting truth, so long as the consequences of the errors are acceptable to the lasting members of the species.

Inevitably, a better truth reveals a prior mistake and rethinking, even though this new truth is just as ephemeral as the preceding one, another walking stone drawing an endless path. In an equivalent way, through the errors of life's advancement during millennia of turning the experiential solutions of a shifting world into genetic-compressed memory, undoubtedly wrought with innumerable errors of trial and correction survived and suffered, each species contours its own path towards a higher truth.

–

Imagining the unconscious, from a dominantly cognizant perspective, as the networked totality of a life energy, efficient, inspired, and capable within extremely dangerous contexts, forced to attain far-reaching, multidimensional complexities of learning and dual self-expression of that complexity through the regulated conscious facet of awareness; inevitably, if unchecked or released, the environment will become fully consumed by this dominant branded awareness and its methods of perceptive construction, so that traits and talents of reality experienced in other species inevitably become side-effects, nonessential to survival, and thereby waste or a nonconsciousness out of stability or connection with a newly comprehended reality.

With the loss of the natural environment and its emotive instincts, technology becomes extinct as a further solution and higher self-consciousness regresses as a dead-end awaiting energized organismal shapes capable of more fully utilizing its use to preserve this synchronized balance of reality's forces.

Somewhere along the way, while traversing a current world, the idea arose to have to transcend this world for another world, without recognizing humankind transcending now within the present state of the world. The best such an idealized awareness can hope for is to reclaim the natural world before it becomes fully lost in deadened antiquity, predictably extinguishing the previously raised sensitivity within the world due to the aftermath in which a lessened environment creates lessened affordance for thought to further inspire upcoming sentient minds with niche innovations.

In this way, the thoughtful realization of humanity has peaked and is in a regressive phase without new environments, or better ideas stemming from the current environments, to stimulate it; or the utilitarian simulated world it might intend as replacement and fitting, deadened end to its evolution and the extent of its best solutions.

–

As a highly conscious species emerging from the forest, deserts, and mountains for the first time, humanity was in possession of a raging landscape empty of all the possible paths its evolving bodily structure could enact or permit it to perform as a successful agency within an environment it has begun to realize. All within its afforded measure of opportunity lie ahead of it, though it can never appreciate until much later the appalling and torturous path it will often track from that original emergence. It required from this starting point, slowly and usefully, to evolve the boundaries of a new, vast, and all-knowing terrain for the mind and its accompanying body: with its subsequent human history reflective in a present-day peak of scarring successes upon the collective human psyche.

With the written word, prose speaks to humanity; poetry, if not simple condensed prose, can speak to the universe. The universe does not need long-winded metaphors of its deeply experienced truth.

–

It is coercively flimsy when a sentient awareness, without the true insight and only the base, manipulating intention, tries to predict the far-reaching ends of its actions by whatever means. This consciousness at whatever stage of advancement maintains a willful separation from reality.

We act from outside and often against our environment: not even the natural environment, but the environment of existence and the potential for a relationship with anything other than ourselves and our vast imaginings of ourselves.

Instead, we act on a natural suspicion towards existence resulting from a lack of empathic understanding of placement and a stubborn departure from anything much more actual than ourselves; and our superficially original innovations of material occurrences, which do not truly belong to us but in conjunction with the unifying environment of which we belong and tend to mostly overlook as we jeeringly mimic it's observed successes as our own.

—

Thoughts are spontaneous symbols hiddenly embedded within the environment and all of reality including the mind. Nature contains symbols and cues for all species to find and observe, like words needing an intentionally reduced focus or similar wavelength to uncover or sight as meaningful.

Truth lies outside the current dynamic of human cognizance. Evolution does not exist outside a conscious mind's appreciation. Instead, it is sequential advancement: it just occurs like a universe without obligating inspection. Nothing within it that detachedly watches, nothing wanting to pull itself free, cutting the tether of its restrictions, needing to peer from its inner being outward to witness an image observed back in acknowledgement. The prides of consciousness are the aggrieved impulses evident of nature's pounding exit into eternity.

Animals are unavoidably consciously sensitive; humanity just usually does not know how to observe it very well. We have a difficult enough time trying to find it in ourselves. Consciousness is not the most important anyway, only existing much like the non-reflective surface of a black hole with all that absurd enormity and hollow mystery condensed inside its unobservable shell.

It is a willful, selected lack of appreciation for reality and what propagates beneath its obvious surface which makes human realization especially unique. Among few, its memory is a greatest strength, not only in discovery but also encompassed within earnest moments of sighted appreciation during which reality smiles back in recognition across vast temporal scopes.

–

The instinctive-urging unconscious will often conflictedly challenge the conscious-intending mind opportunistically using available affordances within reality as an arena, thereby leading it into potentially erratic occurrences and dangerous circumstances beyond its comfort of contention as furthering outcome. As a result of these contests, it will suffer possible severe penalties from its subsequent willful choices.

It can appear as if the urging tendency does not recognize repercussions at the consciously tended level, and the dangers these lessons might earn in reply due to external forces not responsive to its transitional, involved meddling destined as spirited journey. The conscious mind, punished by the unaware impulses of its unconsciously inclined landscape, acts upon its mindful, escapist world. The conflict usually arises due to the cognizant mind's insistence its separate, repeating environment is better destined than that of the other silently unconscious environment's encouraging, venerable nature, rather than being a collaboration of the same world met in conjoint marvel.

–

Memory is the vigorous rousing of imitable past events. The enacted events easily reproduce from the bulk of past experiences and recent observations without having to keep them as specifically stored sensual replicas, for only the context of the moment recalled needs expressive retention.

Memories are recreations more than recollections, and as inventions are often open to reinterpretation and alteration mostly due to causes from the experiences which have transpired since the originating event, and the changes these new experiences have enacted upon the psyche involved. This occurs among individuals and the entirety of a species as a historical context for the exposition of an exaggerated present state.

Consciousness, as with the unconscious tendency, is a sensory communal within reality, the regulated interaction between existence and individual being in all its contortions. Such conscious, attentive focus has the potential to be a wonderous, solution-filled property, if permitted to accept conjunction with the natural facet as a willful, instinctive truthfulness.

Opportunistically, individual perception is vulnerable to outside awareness and the likelihood of mindful disobediences enacted externally as an influencing intimidation. In an essential way, these conflicts of wavering consciousness, like predation of species, will often create impassioned responses between differing perspectives of predatory reality; evolving its most far-reaching, insightful, concepted solutions as reward.

–

Prehistoric art exposes the beginnings of conscious believability seeking a safe, unconscious imprint of its new milieu, like homage to the spirit of thoughtfulness located in nature with the dexterity to first begin marking its simplicity, as a desire for retention of shared memory, to make internally disclosed experiences recognizable in image for a connecting collaboration sought equally between the artist's growing pragmatic mind and the beckoning, contoured stone with descriptions sighted and unveiled within its surfaces.

Original art on stone is a thinly pressed brink between the instinctive world and the artist's quest for the truly artificial world of reality. This is where mindfulness can lead, through newly arising associations, to separate one world into two worlds, since the stone surface is dually a facet of the artist's latent unconscious mind mutually revealed to its conscious counterpart unmasked through the telling fusion of intended markings with intended stone silhouettes.

In the image of an ancient bear, the unconscious intensity of nature has no need for a colored image on a surface for what it sensitizes through the environment or the stone wall as it exists devoid of markings. It is the conscious mind which needs the images to further its newfound resources of awareness as thought bound and symbolized between the observer and its origin, like scavenged stone and other objects of nature shaped into tools or worn as embellishments: an early tribute to a pondering observer of a freshly surfaced reality.

These generations spanning short millennia, before excessive ambitions of context took sway, as the closest humanity's interpretation of a conscious experience came to live in solicitous harmony with a natural, celestial environment, were later abstracted, dissected, and strangely overlooked as worthy origin.

–

The advance of consciousness by predetermined choices in each species inhibits according to its bodily method and continuing niche outcome. Each choice requires energy, and often a choice results in its conservation as a best solution, such as utilizing the environment as a cue for memory and the organization of associated unravelling like the processes of thought.

It would be a strange kind of world if all species, regardless of bodily system, grew to equal or similar advanced consciousness all over the planet, making success too chaotic to evolve among such disparate consciences and perspectives of all those individuals among all those species in conflicted outlooks and contended values. Such an ill-conceived world would quickly diminish into an extreme panorama of viewpoints, a thinkable world inequal according to standpoint, abstract beliefs, and very much an original world like the current state of a species starved into a singular human mindset detached from the multiplicity of nature.

The advancement of conscious thinking goes against the conservation of energy within a species as a better solution, since the mind contributes to increased energy requirements. To offset this gain, the consumption of its thinking mind must feed through successes afforded in response by the environment or the re-shaping of streamlined efficiency in its mind-body relationship with the world.

Innovative patterns of thinking and bodily practice can afford a species the ability to utilize and manipulate objects and concepts within the environment of nature to do what bodily dexterity, niche commitment, or simply fangs and teeth can achieve in another predator species having them within a repertoire of traits: and, from the perspective of this presumed thinking mind, without suffering the restrictive limitations of these traits on its bodily shape and its adaptable skills, thereby multidimensionally enhancing its ability while reciprocally assuming the risks of increasing aptitudes associated with such freely, willfully selected enaction within the full scope of a world augmented by undiscovered penalties beyond current prediction.

Consciousness is the coordination of bodily facets in relationship with its environment. The excess training of mind applicable to thinking relates to subjective human interactions and shared aspirations of breakable human external organizations, delimiting the native environmental influence as an abandonment in the case of most current humanity, or as an unavoidable obligation of colliding survivals.

Humanity may silently learn to resist the myriad ways it could choose to become instead due to the ingrained rationale of the past colluding with the present perspectives of reality as if an existing multiverse of co-happening occurrences, without recognizing the dire impacts prior and current misguided milieus of human activities have had on the emotive thinking and fearful beliefs of its human and nonhuman inhabitants in a present condition.

Proof of a self-conscious awareness might be our ability to fittingly reflect and interpret ourselves, or with what honest depth of actual experience; not just as individuals but as our history as a species possessing its implications on ancient occurrences furthering current validations. Just as the physical comes at the expense of the spiritual, so does the conscious come at the expense of the unconscious. The highest proclamations of truth result from these conflicting acts of equilibrium, like the symmetrical forces and fields emitting through coalescence this material entanglement of particles everywhere we try to look or measure.

–

The unconscious tendency doesn't merely or solely feature itself through dreams or semi-clear, hallucinatory states, but through the everyday as readily as quixotic extremes of events; mostly through utilization of involvement by its adapted perceptions confined through its relationship with a niched truth; due to its unique perspective shared and often influenced by other conniving constructions and deconstructions of the collective, localized world.

Each species creates its unique dimensionalities within a collective milieu, though it might selectively appear as one world. There are much more than a meagre three or four dimensions of reality occurring within temporal spaces; due to relative perspectives, the universe can be inestimably multidivisional in time and space, rather than singular everywhere at once. Reality produces a dream-like state for those capable of reading their experiences rather than simply overlooking the embedded symbols and visions of reassurance afforded the natural world outside dreams.

–

It is not instinctual drives that necessitate violence, it is the deliberate conscious drives we conceptualize and choose to emotionally manipulate that turn violent. Attempts at absolute power are a premeditated act, not an instinctual, niche-required one. It is the silent and varied impulses further inside shaping our necessity to survive which exposes itself through our cognizant realities; human realities shape and invent according to these earliest, unaffected impulses unreasonably accessible as a continuing affordance beyond mere instinctive, subsistent coordination with a familiar environment.

A convergent human consciousness sprung from the original unconscious tendency of nature over great distances, lineages, and spans of time.

Inevitably, to become a successful trait of evolutionary survival, an advanced consciousness must come to recognize itself, thereby accepting responsibility for the exhaustive choices of its species in relationship with the world around it as truer context; reciprocally avoiding expensive amounts of lost energy incessantly trying to find solutions for its proficiently evolved deniability.

Each posture of consciousness is as much a reality of nature as of the individual of nature it expresses, and there is a duality of involvement there: of *ownership* of consciousness either way, even when this ownership is resisted or denied, which, normally among species, is not the case for reasons of confinement, whereby each flip of this duality benefits singularly and mutually without need for excessive exaggeration.

We create ourselves mostly from our environmental connections, establishing enough identity among innumerable groups and symbols of beliefs. With inevitable, rising complexity in a natural space and time to grow without excessive interruption, the successful environment forces multifaceted distinctiveness broadly, redundantly widespread.

Bestowed with the restraints of its being, each new uniqueness, or offspringing species: the universe is an empty vessel for its awareness to evolve into something original within its environmental array of biodiversity.

–

Even in its uncertainties, science impresses equal or more spirituality than religion, yet ignorance of it abounds like the words and depth of true religious enthusiasm. Since both are unequally based on limited knowledge and best guesswork, they are, perspective-wise, attempts at empathic descriptions of an entangled, evolved and furtively closed universe.

Now, our immediate consciousness learns not from the language of a native environment but the language of our conceived environment, and even an embedded virtual environment, though each experience through the same building blocks of reality. Not all awareness exists in combination with a single manipulating bodily shape.

For a deliberate consciousness to grow to evolve a model world so far removed from anything true or meaningful in combination with nature. With agreement or balance, the unconscious-conscious duality can be potent within a species, and especially, individuals within species. This potency depends upon the seriousness of its experience and the range of influence any novel solutions may grant as an advantage to the first to overcome an infringement: the first to find new meaning or truth embedded within the possible image of reality.

Once an inclined conscious species recognizes the true absurdity of its reflection, it is the first true measure and observance of the universe. Only then is a species given the choice to wisely re-emerge from its exile or not, diagnosing and embracing the dichotomy of its nature.

–

Strangely, referencing free will, the unconscious genius of nature, contrasted against the recent surfacing of humanity as a singular force atop the world, is not a thoughtful part of this freedom of will while it is an overlooked, essential component. The shapely conscious mind has its newfound science, arts, or religion in its favor, but the unconscious tendency has eons of felt and nurtured transformation at its disposal. The predisposed unconscious tendency of nature has anciently experienced everything prior to the human conscious mind's evolution into an array of emotive and multifaceted traits for survival. Consciousness is simply one aspect of the wholeness exuded by the unconscious propensity of nature and its potential for change and invention within reality if recognized and utilized within its diversity of inhabitants.

Until the few most recent millennia, this equilibrium of consciously driven organism and unconsciously driven environment, as intending mind, and shaped body, naturally uncovered and yielded within the world. This variety and balancing of traits with forces was how the world incredibly learned to long reflect itself for immense spans of impressive successes: amazingly it was achieved without a true, thoughtfully-dominant, seemingly advanced and disobedient consciousness within its vast collection of species staking its claim until the recent present measured over short millennia and the exit of humankind from its textured caves, mountains, waters and forests all brimming with the thriving, colorful and vocal revelation of life's succeeding, willful force.

3. This Larger, Mostly Unobserved World

Imagining the arrival of light as source for appearance: a newfound property of a simple, wave-like transference of energy. An original, translating birth of sight blinking as a first blinding hint of the evolvable use of this peripheral, unbending residual force, and through physical resource of energy absorption capturing an inner presence of the outside world. As a stirring origination, an organism was able to perceive the incorporeal force for the first instance for evolutionary success as future convergent selection. From there, the wave of opportunity illustrated forward and continuously intertwined with all inevitable and resulting interpretations of it.

The forces procuring life through the journeyings of evolution snagged this observation from the workings of the universe formerly unobserved and readily flooded, so that it was able to imagine local events as they occurred relative to themselves, inevitably through practice and skill over great distances, times, and clarities of use. The forces of life assumed a multilayered trait discovered previously existing fundamentally abundant within the universe, thereby favorably shaping the focusing of a life forced across countless, varied species and millennia of investigation.

Light is one of varied associating forces, and in the case of the universe connecting with life within it the vigor exploited for its diverse transfer of information into knowledge. Without light as an observable presence, crossing an evolutionary threshold making it apparent as an abounding trait to living entities, the forces compelling life as a context of the universe would have little method of decision through any focused reference of attention: everything that existed would exist in abundant, shadowy, unnoticed isolation. Without the observation of light, there is limited probability for a materially extended reality of the universe to venture through and find as the forcefulness of life without an alternative, unimagined, ready method of exchange equally useful.

The wave reflected from the first sentient and physical manifestation of the awareness of light's brightness by life can be imagined, in a current state far removed from its origin, still lingering, radiating its waving arc within the universe at a fixed rate of speed awaiting an unsuspecting observation to absorb a fragment of its blurred and trivial energy.

The instant a first realization observes light, a previously constant radiation of residual energy, as perception, all light everywhere, having already ranged the universe, instantly becomes perception-based, observable information imbued with new meaning for the first instance ready to be travelled, and in so doing permitting inevitable, reaching scrutiny of a surrounding, all-encompassing new universe.

—

A lack of better understanding of the universe could be a singular lack of evolutionary trait to perceive it better due to limits imposed by development being tightly bound to the niched, species-made local environments it proceeded alongside.

A universe does not freely create itself in forceful enlargement from a prior nothingness on a whim. As unbroken change, it could be the universe cannot fully identify from wistful techniques, though there can be a rare confidence enough to know there is no certainty enough to know, with everything in a constant state of sudden occurrences about to ensue. Being the framework for an ever-changing state, the universe is faithfully constant and never mutually agreed on what is about to happen due to an indeterminable predictability of time and space.

When we mean to measure or observe, it is never the whole universe we quantify or see, but merely a focused state, usually ourselves, an object or nature: whatever sensed or thought about. Each measurement and perception of a locality of the universe has an extended range allowing for further measurements and perceptions in conjunction.

With an observer trying to determine or define what something *is*, even if beyond explanation, the search can often hide the reality of what that same something *is* as an occurred experience to the observer beyond the process of definition; as if the observer, by the act of observing and inspecting, is somehow detached from the seemingly external, unfolding explanation it invents to appease a lack of insight or a missing familiarity.

–

Gravity is a relative, weak force in the universe yet can be materially powerful in the connectedness of its aggregated expanses. For a human being, it can become immediately controlling with its seizing embrace, but this is an illusion due to perspective and a lack of awareness of just how small or particle-like a human being really is, especially over the relative, short distances present on surfaces such as planets: for example, a mountain. Connected to proper scale, a human being is minute to the point of near nonexistence on a universal level.

Victim to a calibrated spectrum of proportionality, the entirety of humankind associates the relative burden of emotive thoughts and dramatic events within near-to-nonexistent particle-like participants. Also, at this scale, it is still apparently functional: as one particle inside the vast empty space or defined probability of one of one hundred trillion atoms inside each of the one hundred trillion cells of a human being.

On this scale, a human being should have consideration as vastly less substantial than just one particle inside one atom of one cell of its body in comparison to its entire body relative to the universe. Yet, this human being would apply no consciousness or willful substance to a single particle within an atom, or massive groupings of atoms made of such particles, each with its apparent, probabilistic nature constructed of the energy of unused potential: fulfilled and actualized within a physical reality through contact and exchange of energy and information.

We occur as one of innumerable fluctuating probabilities, finitely moveable within a universe raised through the breakage of limitless bound points of being. Imposed limits are based on predetermined choices. Reality becomes ever more contained by choices and determined to those selections locally while retaining abundant inventiveness within the range of new potentiality emerging through these limits.

The locality chooses its world and the ensuing consequences through the product expressed by its existing evolutionary culmination of selections and the subsequent plausible outcome cast through the interactions of its available forces, in cases including divergences into species.

–

Gravity, an essential facet in the behaviorism of the universe and its range of potentialities, behaves as an act of balance just as heat attracts single-directionally to cold and its achievement of equilibrium through this method of fluxing one-way attraction between networked objects. In this interaction, the matter particles of reality are as much like the allocation quality of temperature as they are the physical objects needed as medium for the distribution to take place. The universe exposes itself through such vigorous presentations of single-minded, symmetrical attractions between all localized and related objects according to qualities each contain and the mutable forces each convergently exude.

The universe is a complicated place. It required reduction of extreme states of probability down to the level of the finite meant to construct compressed realities made of the same reduction of probabilities: within new environments empowered by the forces, within and atop the surfaces constructing a reality, the previous breakage had established as a prior result.

To materially exist, something must become measured through contact, and even sensory interaction such as observation can excite this manifestation. It is the same with textured physical reality as it is with cognizant reality. A thought, or the evolution of a thought, must be reduced and measured to become real enough to hold it, since thoughts are characteristically easy to lose track of and the reason for language, archetype, art, science or history, in the vastness of that articulation and all its implications: to forcefully render the concretion of our thoughts through association within the environments of reality much like the drag of a field will materialize particles; to symbolize something so we didn't have to keep re-finding it; to impress our thoughts upon reality and make them physical so we receive and find the impressions back from the surrounding world.

Except, humankind is a celestial work of fiction, as these are not exclusively its thoughts lavishly imagined. The thoughts are an invention of nature as survival, as adaptability, and the fixed symbols are already there and believable, striving through an infinite network of heroic presentations of life towards solutions for the furtherance of fueling evolution's instinctive designs.

Naturally, thinking is an adjustment of the environment. Even trying to better it, which is self-defacing since thought archetypally and individually contains within the environment. All life and consciousness in the universe have little to no context without a niched environment to sustain and fulfill it.

—

Without determination, there could be no further unpredictability in the universe; or at least none with lasting significance. The prerequisites of the universe evolved its forces and particles into skillful energy embodied as matter, including subsequent cells, a galaxy from the viewpoint of the particle to the cell, and eventually the cell to the organism, and prophetically the organism freshly standing before the universe with an absorbing stare.

The particles of the universe had to organize before the cell and before species. The particles, being a universally seized solution and context for evolution, had to cultivate beforehand into the fundamental particles of the cosmos, and all its manifold experience of mutual exchange into workable stability as a strict conduct for celestial evolution. Inevitably, the cell evolved into a bloating of insignificant awareness intended to resist entropic chaos through riding the forceful waves of a delicately tuned universe, feeding off the provisioning energy and empowering surfaces of a small planet.

This might seem simplistic as a description of a fundamental world: if not for the necessity of particles evolving from unfilled potentiality into the meaningful substances for life before the arising of a single cell. At such scale and volume, the particles, an obviously vital unit, *became* the universe for the perpetuation of each subsequent cell and organism's evolvement.

From the universe certain freedoms of self-sufficiency arise, inner origins presented for actuality to use through the spectrums of temporal equilibriums and forceful symmetries; yet embedded within that freedom lie paths and methods of suppression of these ascending choices just as validly presented in complex, measured systems. Limits impose universal capacities however extreme, mutually enthusiastic and satiated in experience for the necessities of harmonizing the permanency of an inestimable reality.

If there is affordance and will for it, the universe is the embodiment of itself; eventually, things, allegedly side-effect and meaningless believability such as species, permit the opportunity to range the fields of experience. Each species must go through the same, often long-winded, and horrible evolution for the chance to become successfully contained within a caging niche.

As with the conjoined evolution of species atop a shared environment of the same base surface, it does not evolve true solutions until much later in its diversity, once all the ranges of extremes have exhausted in extensive, potent repetition of reexperience as the means of refining emotive and sensory responses. At other times, it never comes to a complete solution before an opportunity expires.

The underlying universe and all physical nature are the crests of technology and there is no significant other contained within it as the selected solution. By choice of existence, reality assumes its consequences as it strives towards an evolved conclusion according to a surviving premise that a consequence is not solely answerable for its origin.

–

Each measure of existence exhibits a world unique to itself, with its own properties, traits, behaviors of being specific to its context or moment; each interpretation occupies the same existence of unique constrictions and reduced, finite conclusions. An atom is only ever an atom, and a molecule only a molecule; but in entangled combination confirms higher dimensionality of opportunity which evolved, perceiving minds can easily assume as thinly surfaced three or four-dimensional space.

Reality possesses dimensionality only limited by the potentialities of the local moment encompassing all perspectives past, present and predicted within its range of sensory influence alongside the consequences of innumerable overlapping others probing the same dimensionalities. Each instant of reality – with each uneven moment everywhere unique to itself as a unit of perspective – contains infinite dimensionality expressed through infinite sources of entangled material and immaterial being behaving in illusive, symmetrical coincidence.

Reality is in a state of ever-changing, manifold superposition or multiplicity in which extremely low probability occurrences can sometimes fitfully occur. The world is flexing construction, the measuring of probabilities, the breaking of uncertainties, and the synchronized, ever-occurring revelation of infinite choices and forced repercussions as a current culmination of the past.

With mere energy and degrees of density, the universe exists as configured systems of silent stillness. In a universe, devoid of life or consciousness, without any means to perceive, there is no need for appearance. Light becomes radiating energy without association as an emotive source for advantaging inspirations or having any rooted obligation of its wave-like nature rousing to imagination.

The tendencies of evolution teach life to perceive energy and force as a solution to its dilemma. This occurs with all of life's utilized sensations. Life, in this way, furthers its own perceptions upon reality in unique illustrations through evolutionary benefits. It is the observer perceiving this light as wave and length to give it a quality relative to our perception of it in conjunction with all other awareness, experience, or measurement of it as origin and placement.

–

The universe predicts to be deterministic up until the present, moving moment and beyond. It is continually, single-mindedly made, much as with the future which is open to interpretation fully based on the fluctuations of its accumulated past. The universe is whole and complete, following predetermined solutions or laws established long before. It has an inevitable past and future with only the variations of the present to be determined; but the result fixates regardless of how the passing instabilities turn out. Nothing can alter its path, or at least it visualizes this way.

Lessening things down to the experiences of human interactions, humankind is, unlike the universe in this scenario, undetermined by present or future beyond the whole, natural succession of tendencies resulting from its past. Enhanced with an ability to cogitate and predict, species shed a deterministic nature confined atop a planet.

Nothing within these human vibrations upon reality can in any way alter the deterministic nature of the universe – a deterministic nature to continue its forward evolution – as they are much too insignificant to ever factor in: humanity's greatest involvements will have no bearing on the universe's outcome. In this way, even if the universe were determined through its forces and evolution, rather than balancing atop a swaying equilibrium, temporal species curled up atop planetary waves of reality are not.

There is no real separation for humanity co-joined with this reality, or at the least an immensely constricted, erratically occurring likeness to it. The planet's evolution is as the evolution of the universe as any countless origination of its dimensions along any current state of its making.

–

Life promotes the organization of energy. It confines, gathers, and transports energy. Without life, there is much less efficient manipulating motion of energy within a broader range of overlapping local environments. In other places of the universe, such as extreme structures of nature like a star, life excludes organizing energy through any viable enough means.

On a stable planetary surface, naturally occurring constraints organize energy into repeatedly flowing systems. Once energy takes on shape, which it must do to pronounce its occurrence, it strives for balanced efficiency along whatever path most successful: one method being life wherever it is probable for success. This is life that is not simply conscious of everything: it is perceptive of itself while setting about the task of further evolving its efficiency of organizing the flow of energy along surface thresholds. But, during proper conditions, this consciousness can evolve too: into an initially powerless recognition of its role in an intended body of actions embodied in this practice of creating and furthering a reality.

–

Everything material in the universe, from the smallest dust particle and inanimate objects to conscious actuality, holds at least an amount of information and energy for the universe. This information, as immense as it is, is a small, raised portion of more. Our reality floats on the surface of this larger, mostly unobserved world.

Particles acting on observation does not exclude the possibility of them not simply nakedly revealed but acting in revealing correction with observation, with particles and a sighted, sensory mind being the environmental context for reality.

Since the particles of the body become attuned through the action of contact or sensation, the bodies act on fiery sensitivities encouraging the will to further instinctive actions, equal to how the contact occurs in return. From the particle's perspective, the seeking observer equally appears for it in coincidence with it appearing for the seeker.

Isolating particles in experimental objects creates an unnatural world for a particle to reveal its true nature of wave, field, loop, string, or whichever-like blending within its native locality otherwise so easily detected everywhere else an observer looks merited to a common, evolutionary benefit.

Quantum fluctuations and dualities define the alignment of the present. The present forcefully chooses or determines everywhere. Even at the level of the quantum, all is not certain, as there can be correction, delay, and adjustments made in accordance with an ever-changing, multifaceted explanation of the present inside the spaces of all moments; with connectivity and awareness as a dual-sided will, and the moment completed through the mutual choices of each dissecting dimension. In instances, when compelled to due to an absence of will, reality can make the choice for betterment and dissolution.

When an observer looks at a particle, it can appear still, but even in that stillness performed while looking, it continues as a wave no less than before, even if the observer somehow never looks away or preserves contact. Then, the observer witnesses a string permitting a continuous flow of reality, or the wave in attuned motion alongside the observer persevering the same, so long as the observation upholds. Observation is a method of interpretation. Reality's entanglement is a vastly dimensioned wave, as valid and evolving as the environment it ensnares through its workings and fabrics. As a discovery, an entangled world should be an expectation and unsurprising even while not exactly fathomed by reciprocally entangled minds.

With particles acting as waves, and forced to select, or sequential particles acting and performing the repeating wave sequentially forced to make sequential selections, it can measure and thereby make one choice or another, even seemingly going back to choose and proven by the observations forcing its occurrence.

Same as the alternatives, the choice has occurred already as part of the wave. Choices waiting to happen and when presented with choice, a particle enacts its natural wave solution. From here can arise ideas of a multiverse of branching, expanding choices, but there is no one or the other, they are both susceptible to multiple viewpoints and separate temporal points of view. They simply all belong to the same universe without separation. Once contact stops, it ceases to occur, unless observed anew, consequential to rediscovery.

It is the observer expecting the particle to perform to illuminate its natively unseen state, in a world where it must do *something or nothing*, thereby expectantly exposing it in its unnaturalness and without proper context; committing the particle to an original choice it must then assume as a trait of its further evolution in wave-like unison with other, momentarily absent particles other than the ones measuring and observing it.

A particle isolated without measurement or observation exists without change, timelessly anticipating the next selected wave to ride. In such a state, its material reality ceases to exist: to observe it in isolation is to unnaturally observe its feedback in a quest to find its true behavior, or to have it spark and play tricks for a better lack of understanding.

–

The only plausible infinite is an extreme of probability: and that is unlikely without the infinite of opportunities available for its occurrence to happen. It requires absence, void, and does not completely begin.

The only infinite is nothingness, exceedingly rare as it impossibly stretches forever as an extreme feat of probability. Then, once initiated, once something comes into being, from the first apparent lasting vibration struck by reality, chance originates, and infinity is abruptly over before it began. This new world is finite and constrained by its own cumulative inventiveness and workable evolution. The vibrations finitely stretch. The nothingness of infinity does not need to exist where it is not useful to a new finite reality.

The backdrop of the universe, of conceptions of space and time, considers a wave of unannounced probability: a field of nothingness until something occurs to remove it; and an ancient, obsolete infinity absent from the beginning.

To travel through apparent empty space, such as a hurtling rock, planet, or any entangled energy with trajectory, is to travel as a subjective event coursing through a vast stilled ocean of empty, awaiting opportunity. All that exists materially is the travelling local event and any fading effects, dispersing by whatever limited force possessed in its wake.

If capable of being measured, long travelled, radiating light can be absorbed or observed from anywhere space is considered empty, cold and dark: with the potential of sighted unification of past and present shared over millions or billions of earth-year spins from emergence to echoing absorption, with renewed reflection and expanding wave trajectories of its subsequent event.

On the scale of essentiality, the physical and nonphysical worlds are the same thing. The expense to create a thought and the expense to make it physical involve distinct aspects of the same energy and forces previously evolved within the universe. It is the receiver interpreting or utilizing the energy which translates it into an object the same either way except by its configuration.

The light which reveals the world to a physical observer is nonphysical, since, travelling constantly at the thresholding speed of light, it carries with it no corporeal substance. Yet the physical recipient, being a potential evolved entity within a reality embedded with fixed, regulated constants such as light, does not necessitate the observer having the capacity of realizing all the potential solutions of either the light itself or the universe containing its force, due simply to the requirements of survival as a continuing species atop a planet surface within a trivial space of that entire existence.

In the likelihood of the universe retaining information, it should certainly know what we know, individually and as a grouping of species on a planet, as part of its evolution of everything else anywhere else in a small, compacted locale of curled up, hidden reality.

There are likely to be multiple explanations for the universe, until eventually all the countless recesses of possibility have exhausted and perpetrated and there is nothing original to newly postulate, and with each remaining viewpoint attaches multifaceted explanations and measures of beliefs without dominant certitude or reduced truth.

—

For a world to cohesively function it should not seem remarkable the universe, and undoubtedly life, could not exist without in accordance with some formula of a tangled, coupled state: at least permitting it to occasionally execute the impossible while remaining within the confines of reality's forces exerting a will upon it as command. Yet this entanglement, once begun to be disclosed, will appear surprising and unbelievable to an eventually consciously thoughtful entity first alerted to its hints no differently than its enmeshed history reacted with surprise and disbelief at countless newly worded observances of the world as they were first revealed without much deviation in mass response to anything similarly newfound.

Humankind manages to penetrate the history of the cosmos as far as light, a solitary connecting force among the observable universe, can present to it a past relic through a delicate evolutionary exchange, with everything within reality's containment in a never-ending state of absorption of old information and reflection of current information.

Nature does not mimic technology; human technology crudely mimics nature, especially lacking in its sensitivity and malleability. Though extremely separated in advancement, the impulse is negligibly the same. The cells of life, a rough trillion in number for most animals or plants, sustain by the reoccurring filtered consumption of fundamental quantum energy.

Events within local realities can persuade the past to alter by new knowledge of the locality or system gathered in the future. Even in science, belief, and everyday life, we create our perceptions in our outcomes. We simultaneously cause, reason, and measure our own reality as it evolves.

—

In extreme events, there does occur the otherwise impossible situation, particles cannot withstand the excitement and cease unless constancy returns. In that moment, a nonreality contains, fixed in place without change: the compressed innards of a black hole are compressed probability held in check as effect of extremely improbable, evolved forces. The extreme event is the highest excitement. Yet the lowest excitement, the absent in scale, is an extreme event as well for its boundary or measure of containment of events, even if conveyed as stillness through an animated, behavioral, and shapely exhibition atop a planet's surface.

Since a particle exists in superposition entangled by past, present, and predicted states, it can measure as a particle while retaining its wave-like pattern. Since all positions exist simultaneously, it can be within the countless realities there are to measurably observe it. In this way a communal multiverse may conceive existing in all probable positions based on the existing locality's entanglement with each particle, creating all perspectives of reality in conjunction. This is solution of the universe life adopted to survive its subsequent environment.

What exists to an observer as a *moment*, continues – to another unobserving observer – as a wave of multidimensional, defined probability. Yet this unobserving observer contains its own *moment*, exclusive from the original observer, whose reality is part of a wave of announced outcomes to any other later observers not taking notice right then: with its lateness defining its spatial difference.

Everything, encompassing all, a past and present state, is the consumption of equilibriums. All human and worldly endeavors are the exploration and documentation of probable states to the furtherance of new probable states. The universe is the concretion of probabilities as backdrop to new probable explorations.

The universe mutually reveals itself to humankind as it delves to be by the unifying perspective of useful detectible beliefs, with each perspective an individual, kept truth among an immeasurable multiplicity of truths within reality.

–

Humankind happens within limits of thought it exists in only small, simple dimensions. There are connecting and facetted dimensions apparent in ordinary and challenged interactions; these dimensions result from the mind-body and the attained skills acquired through its relationship within its involved environment. Each mind-body excludes from these dimensions equal to the extent of its unique rapport, practiced learning, and willingness to fathom and invent within this multidimensional plasticity. Due to complexities, most will specialize within facets, finding limited representations within these deeply dimensional realities. Often, the pursuing goal of mind is to minimize these haunting and exhausting multifaceted realities into a practical, niche method, just as species evolve within the confines of a selected locality.

Much as with the spectrum of light, a mind can consider existing within specific wavelengths when absorbing reality according to the current state of its evolutionary transition and prior experiences. Light, a transfer of force and energy – without which there could not be a reality as understood by such awareness utilized through an affordability of a type of radiating energy itself mostly benign in the universe – could be said to not contain a wave unto itself but only the one applied by the observer. Through the act of perception and absorption by the body and mind into the impression of a visual occurrence, the observer assembles ongoing change applied by the wave-like volume of light. The light goes in all directions, and the waves overlap and bypass in such a way that from each perspective the waves of reality available through light are a completely singular wave to everyone, and even more like an engulfing than a wave. Through interactions with other waves, information and impressions continually absorb and emit within and upon reality, stretching through the locality and beyond even into the minds of its inhabitants.

Without the inhabitants, there would be no wave-like quality to light creating appearance; it arrives within reality through an evolutionary trait of life to apply wave-like value to a force coincidentally occupying the same world. Through this incredible feat, from anywhere within its space, light creates the illusion of an existing past roving the present state of reality through the enaction of its reliable, defined waves of emanating energy pronouncing a cosmos for all sufficiently evolved perspectives to glean individually and collectively.

4. Past Curled Inside the Present

The world is much more and different than we habitually learn and teach ourselves. Mostly, like a stranger, we live in fitful exile from the fact we exist placed within an environment, a world, and a universe.

A lack of obligation for what improves and preserves humankind in this world is reciprocally immense. As a species choosing to bravely stand it alone against the intense balances of reality, humanity is answerable for this choice and the reflection it inevitably imposes, foreseeable only dimly or through distant hindsight during worthy moments of individual or group deliberation. Not just those current to humanity are complicit, but generations in the past and into a highly probable future due to this repeating persistence.

There may not be any final answers to existence within an unannounced allotted time. Reality reveals itself through the inherent unevenness of balancing forces. There are always alternative expressions for any reason or seeking. There appears to be no one truth, just a comingling at times resembling one and then fading into illusion. Existing in a vacuum between and within extremes of contradiction, there would appear to be no correct path with certitude but solely through the opportunities of convergent rediscovery among human beings along temporal lines of reality.

If, as with the universe, life is a measure of vigorous equilibrium, then we each reap what we sow, physically and spiritually, according to any true measure of this balance.

There is attunement between life and existence. Reality reveals itself to life in subtle ways not easily recognized; but to the observer keen upon this reality, there can be mutual and meaningfully unique moments of recognition. These moments are not obvious and easily blend with ordinary reality, often going unnoticed, and are not believable, predictable, nor legible to everyone.

–

Stories and occurrences are not all the same as each witness sees events in a particularly unique way and remembers them badly mostly through simplified invention. We rewrite our past according to our conscious bias of reality and our placement within it, which is often misleading and poorly understood since we are often applying a mistakenly interpreted past upon the present through methods that resultantly compound errors when we try to predict the future.

The mind learns by right or wrong choices accumulated through a lifetime; in some cases, shortening it; in other cases, given enough experience, it can become a cumulative predictive insight, especially empowering if it becomes prominent and accumulates within a species or a portion of a species; thereby increasing its likelihood of becoming buried or transgressed in unutterable wonderment, secretive and mystical against a backdrop of the countless everyday scattering of ill-informed choices and overlooked consequences.

Sometimes, among the choices, like walking stones, a path is divulged within reality through presences which have survived excessive and difficult alternatives yet continue to wander and observe with a heightened sense of confidence; and infrequently, a wanderer will reveal a newfound belief for future absorption and evolution by the masses; or if mostly practiced unobserved, the wanderer experience will profoundly fade back into the wave of reality as a singular expression of existence not to be further disclosed unless convergently rediscovered through an individual origination.

The quieter and more controlled the experience emotionally, the better a stranger to the malevolent detachment of humanity through personal experience, the greater its beauty and inspiration become as an equilibrium between mind, body, and reality.

—

Beliefs over-utilize as devices of justification for corrupt behaviors and as righteous cover of fearfulness. A fear of admitting a placement within the moment, of blindly acting limited within a reality with its rules and forces enshrouding it like a fetter against freedom.

Ideas can enhance the fear of accepting a justifiable role within reality's fabric; instead, reaping the easy, unafraid role akin to self-advantage readily widespread and sown upon the world by less-inspiring predecessors.

A path founded solely on the beliefs and ideas of others is the same as following no path at all; and having not existed except as a potential carrier of groundless force or will for others with even less integrity towards the world.

Human history saturates individuals to act on impulses not their own towards a greater good founded on ruse yet believed with conviction to be correct in the moment. Throughout history, these minds profess a dilutional and careless judgment of righteousness. This type of dilution of human reason stretches into dangerous realms of reality and can be an undoing of any cognitive attachment to existence if left unchecked and permitted realization.

Nature easily recognizes such subterfuge as readily as humans can easily ignore careless and improper actions so long as an external source justifies it. Corrupt behaviors are often the result of group consensus, however improbably conceived and lacking in broader understanding, that an action or sequence of actions are just and its consequences acceptable at whatever cost for the endgame of their own security. Only rarely do individuals act discriminatorily in isolation. The prime motivation of this type of grouped pseudo-reasoning is the appeasement of fear and the declaration of resentment, each with well-worn stratagems of manipulation and ploy within the confines of corrupted, self-indulgent minds.

An untrained mind can be a dangerous thing detached from its environment, from reality, living alone, in fear of its impossible nature, and no other conceivable outlet for action except self-service among groupings of like-minded individuals ready to abuse corrupted ideas readily provisioned by history.

–

Action and impression can be so great that the universe pauses to take notice and become affected. Among humans, this admiration usually arrives through a newly uttered passion, insight or great sacrifice; and thereby exposed with the advancement of change to near negligibility due to dilution of experiences, excess disbelief, or lack of sightedness through an ensuing history, leaving the occurrences of great universal feeling to exist as an absent or localized presence within the networks of human interactions: among the loss exists the higher probabilities of the relinquishing of universal movements within the human consciousness due to the erosion of individual inspirations and the mass blending of experiences, thereby leaving it solely to its own devise as a species within an empty universe suddenly devoid of vision to boost its physiological, moral and social evolutions.

–

It predicts that an advancing species cannot truly become civilized until it has fully exploited its natural resources and spent its environment due to a resentment against nature: the strange, uncalled for predicament of having to claim origin of subsistence as a species in such a place as nature.

Any individuals of a species utilizing the emotive consciousness of the environment to unreasonably exploit its own species and corrupt its environment's message should appear cannibalistic in nature, even if by reason of misinterpretation they justify behaviors by promoting the physicality of actions over the spirituality of actions without recognition of its duality.

Balance will always return, even if its perfection only expresses as a relative instant during the full measure of its transition: the greater the imbalance, the stronger the energy of the resulting rebound. For imbalances, such as entities temporally slowed through subsiding on a planet and bound to its local qualities, such transitions can seem unremarkable until experienced in context and enhanced by scale and force beyond the assumption of the imaginable.

–

There are limitless, deep-rooted tragedies fundamental to the world. It is reasonable to be prudent in this life when it comes to a mind's need to protect against and restrict access to new memories; thereby attempting to avoid further elaboration of humanity as it enacts itself within its embedded state of reality. A reality it has rarely been capable of fathoming; and proudly empowered by a lack of understanding, a mind mostly takes no culpability for its actions, merely making of the world a source of vindication for the full, exhausting spectrum of its possibility. A self-aware species needs to exclude itself from the idea of indifference its actions carry no consequence or equilibrium to self, others, and world.

–

When people look for and find meaning in signs and symbols in nature or the coincidences of reality, it is the expression of past and immediate desires, fears, or the ambitions of the subconscious self; and it is often mistakenly confused with something imposed upon the future or exclusive to themselves, rather than the current reflection of a state of being, much like a dream in its karmic messages and interpretations and often just as nightmarish to encounter.

Reality has depths of context. One symbol can have vastly different comprehension dependent upon the aligned perspectives and histories observing it.

One symbol, from differing viewpoints, can appear dissimilar; and with experience, it is the same, one event can view multiply and differently while retaining the singleness of its moment. The single yet multiply encountered outcome is the present consequence of cumulative points of view with each point being involved in the shared consequence while maintaining specifically unique present states of experience. It is a vast enthusiasm stated as a momentary finished, continually ephemeral current state carried away individually within the mental confines of innumerable probable meanings, predicted outcomes and related past influences.

–

Any who practice a religion base it on their beliefs and actions in outward appearance, as if their truer inward reflections were miraculously unknown to the world through seclusion within themselves and concealment by gathering. They would do well to concentrate any self-directed prayers on the hope their god is absent.

Religions, ideologies, systems of belief and responses to them, like everything else, are products of environments. These environments are native results within reality, like restricted behaviorisms of evolution due to localized potentials of related space and surfaces, no less than the imprinted development of evolution cast within reality itself through genetic strategies;

meant, at least in part, to contain the catastrophes risking organisms from a dangerously contended external world.

The atrocities of the emergent human spirit are the consequence of a human dread to find or admit oneself within an existence, and the calculated advantages perpetrated due to this fear. Our fears draw such forces, like predator to prey, capable of subduing entire collectives and destroying peoples as a disbursement of resentment cast as the ebb and flow of dissent; and to find the countenance of violence and extremity so compelling, when it is merely no more than equal to any alternative except in its self-advantaged reinterpretation of its environment.

The misreading is that this reality is not about living atop a planet in an established world; instead, this is about living within an existence, completely included, impassioned by it; everything embedded within and filled with this existence; everything is a manifestation of this existence.

In the end, being an exhibition of reality, there seems little choice but for a stranger to trust and accept it with a measure of pride, though this choice doesn't necessarily extend to other interpretations and responses: to a reality shared with a risky overabundance of beliefs and ideas without much truer understanding of how these exist as simple features of the many facets of reality rather than being inbred and unvarying.

–

Just like with an electron, a moral choice is not singularly culpable for all the other probable choices simply because they exist as possibility. It is the final choice, the materialization of the electron, which determines its true choice in relation to all the other single-minded adoptions of other electrons.

In the same manner as with moral choices, whether acknowledged as moral or not, through perception and measured observation, as a means and method of quantifying a determination, the universe singularly contained within each viewpoint and moment of it: inclusive of choice.

Anything resembling truth is not waiting to find or discover it private to reality, but a thing stirred to belief through the enabling actions of its divergent inhabitants.

–

The dual nature of human consciousness produces ideological evolvement into democratic agendas as compromise, thereby making democratic actions constrained by impulsive future changes.

Democracy, born from revolt consequential to such an acquired nature, lacks certitude in its enaction within the world by attempting to represent opposing views together, thereby hindering overall progress through excessive waste from competing natures opportunistically advancing opposing views. In this way, it is not much different than any other political ideology promoting one singular-appearing viewpoint or its opposite, other than by its methodical attempt at compromise between the two extremes leading to mistrustful, resentful actions in response.

Democracy is the inevitable, precariously popular representation of reality's dual nature as a best answer during a transitioning state of being for its species.

–

We are as we learn ourselves to be equal to our denial of ourselves. There is a better truth than anything pervading human imagination so far. Imagination being our ability to connect and interact within an environment to which we have effectually detached ourselves. We have found the simple comforts afforded by this world. We can mold physical shapes and play with thoughts like leaves swirling in a circling wind. We can pride ourselves on our willingness, invention, self-important beliefs and take recognition for the ensuing actions. But we cravenly lack the ability or willingness to find through our imagination the deeper means to ourselves, to genuine admittance.

Intermittently, this imagination does arise within a present state of being, but based on the present *now*, it doesn't as a rule have lasting influence due to its adjustment and absorption back into the furthering wave of reality to which few scattered traces of these imaginings remain, hidden through dissolution in plain view as a contemptuous homage to our lack of resolve and the erosion of our thoughts and connections with the world.

–

Truth is individual, but this does not mean that individual truths cannot align closely coordinated, just like a statistical range, creating interactive waves of irregular, pointed connections. It is not about truth or even meaning beyond self-invention. It is only what is believable that is real.

The world without beginning or end, without sensitivity or desire for anything real. There must be substance for such feelings to take hold; it must be mutual, the coalescence of entity and environment in poised correlation.

Due to the density of the entangled state, the loudness is in everyone and everything to the point symbols are no longer true and feelings or sensitivity cannot trust. Only the individual remains to remember the magic of experience when played in coincidence with reality and the mutual wonder that it reaps.

It can make you uncomfortable, the ineffable, obscure bigness of it. The near infinity of it in scope and value. The feeling, in such a vast universe, of the absurd smallness and insignificance of each individual thing within it.

Truth exists in the context of what was as a broad reality, separate and before the current influences of life. The recognition of placement in relation to the world, embedded within and part of it, reduced.

This loss is the choices made by prior examinations and the continuing exploit of interests in its current state, leading to an evolutionary birth or unit of inhibited potentiality sullied through the accident of its birth by the means to its end in getting here.

—

The immense and ancient measure of feeling deep within life makes it vulnerable to manipulation and ploy by corrupted wills; not solely by the environment within which it aroused such an essential adaptation to survival, but also by the by-products of the environment: its subsequent evolution of life inclusive of the human species.

Failing to look inward, these scheming human intentions cast no genuine reflection, possessed, and inflicted by a self-absorbed empathy, so take no culpability for their actions except as guise. They are the nemesis of the trickster myth as they exist not as lesson or honest messenger, instead gaining power through group-likened, self-explored sympathies and fear manipulation as they prey upon the deep sensitivities inherent in others unlike themselves.

—

Such a profound capacity among believing minds to overlook coincidences within reality, or to outright dismiss them, when they are so obviously prevalent among an entangled reality: and so saliently meaningful and habitually a joy to detect and encounter.

If you ignore it, it tends to ignore you, and both you and it connect solely through a lack of appreciation; though, in defense, such marks upon reality take determination to achieve through methods of especially rigorous exertions of body and mind meant to re-achieve a prior state of being no longer being experienced and lived to the pith as survivable necessity: a rare proclamation to achieve with the current, highly incoherent state of humanity devoid and rid of such sophistication of mind and body as a function and extension of a native environment.

–

Beliefs and ideas are an effective means of controlling and condoning mass behaviors otherwise intolerable, and the symbioses comes from mutual, self-evolving agenda: or a starving of will or lack of skill in conjunction with reality.

Ideologies, like humble ideas, envision as guide or best speculation from a coupled perspective, rather than simple dogma. Given the choice between a society or culture that needs to arm itself to protect against its own or external threats, or a society willing to accept the potential risk of disarmament as agreement, it could be debatable which society exhibits true strength and which timidity. Though being armed shows great quantifiable strength, for a group of individuals to unarm while dually exposed to the inherent risks for greater cause would seem to exhibit truer mettle.

This abused and confused ideological debate has formed the basis of human history and the dually resulting exploration of this fear of intention has been the cause of its sublimity and horror over millennia. Within nature, through nonhumans and humans, this prolonged and agonizing struggle of collective psyche explores far, vast depths through the critical experiences caused by the processes of evolution on its transitioning life. To witness that depth at a later point looking back is to witness a deepening intent within those vital progresses embedded inside reality in its buried outcomes and the standing consequences of those countless debates presented back as an arguable necessity of reality in any current state.

–

If there is any measure of sincere, generous passion expressed through humanity, beyond us turning our instinctive sensitivity for our environment into something with greater prominence to ourselves, we learned it from nature.

But truth, once spoken or found, can just as easily become absorbed and lost, and usually with purpose for the dangers they can impose upon self-minded experiences. Uncovered truths now lay in dust, ground into wishful non-existence by this massive horror at their implications.

Unfortunately, it is likely that any deep true human love in the world hardly exists since such feeling should influence the direction a species should take over the course of its historical outcome, mostly viewable through its current present state.

–

It should hardly be considered a good thing that with the current successful remaining human species, having eliminated its evolutionary brothers and sisters whenever they encountered them throughout history, and with the lack of isolation and new niche availability atop this planet due to human absorption of its space combined with a shrinking environment from which to spring new concepts, the human species has, as essential, stopped evolving on this planet. Inevitably, it will be the same for all else as a fitting brand of luck embedded within reality for conquering species.

Something absorbed in one part of the environment will reflect from another part, often separated by a deceptive measure of time.

–

The first inklings of human interpreted morals within our consciousness would go back at least as far as the use of stone as tool and object as ornament, and the first stirrings of instruction vibrating within the expanding cognizant yearning for these early native connections.

Later, this beginning attunement evolved into a new consciousness, coupled with the coincidence of a bodily form potentially capable of physically manipulating the environment into newly idealized shapes, with the ability and will to reinforce and expand this brazen new teaching of reality achieved through the environment. Its impressions cast upon walls of caves, objects with color, and inspired an obvious awe in these early experiences of a newfound perspective of the stars, landscapes, inhabitants, and the surfaces of things as they appeared to a newly advanced awareness. It is not hard to imagine a quiet incorruptibility in these first recognitions of truth within a harsh world. It is also not hard to imagine the same world having an eventual corrupting influence on the fresh rising awareness equal to its willingness to follow or fear it.

Inevitably, countless minds have explored this environment for its moral countenance through the recesses of countless books and experiences like an environment's niches explored by organisms until every possibility is found and skilled upon, by extremes of human fasting and deniability, through grand ideas, systems of belief and procedures of behavior all meant through one endeavor or another to find appropriate human reflection for this native enlightenment's interpretation.

This act was first perpetrated by the environment itself as affordances for species to survive, and later, consequentially, the human species became aware of its potential role in the great scheme of the universe's design, or at least a minuscule portion of it atop a planet far from anything locally tangible to direct it otherwise.

Human morality has become coerced to exhaustion into disproportionate methods far removed from any original purpose or intent of the environment it found itself embedded, far from any original awe-like wonder or true faith through selflessness and confidence. It has become wrought by forces incapable of its fuller understanding, due to a resentment at the self-lacking faculty of these forces in comprehending feeling and thought as it pertains to environments, resulting in the surrender to exaggeration and undeserving self-purpose among vast portions of the populace least capable of its lessons. As with democracy, this vast outbreak of understated misunderstanding ends up being the guide for future course of direction, as the majority directing choice has the least understanding of the impact choices have on future outcomes on larger scales.

The entangled connections permeating life and existence can be difficult to tell before they occur: this entangled state exists to express a proximate certainty atop a rolling wave of befalling probabilities amid overlapping localities. It is hard to imagine reality being truly describable as a finished explanation considering the endless circular symmetry of interactions needed for life to be survivable among the forces and particles building up impressions of even a local, present state for countless individuals to affix through evolutionary convergence. It is obviously difficult for the human mind to appreciate the unfathomable simplicity with which reality presents and experiences through vast intricacies of perception, dimension, feeling, and belief.

An advancing conscious mind learning to coexist within a universe can become confounded by the possibilities of its awareness, and unhindered by its implications as easily as persuaded by them while examining ensuing actions and the capabilities of resulting actions. Not all minds find equality in creation, whether it be genetically, through the experiences and skills of learning, ambitions, or physical or mental imbalances due to environmental effects.

The chances for advantage taken against an emerging consciousness within a species emerge through the inequality of understanding among its individuals. This creates the opportunity to insert corrupted ideas into these ill-informed belief systems. Organizations of beliefs and ideas can arise simply through the opportunity of need established by this heedless collective of perpetrators, formed through the potentialities or scenarios due to such a mass inequality of perceptions, without any one individual ever attaining full comprehension or perfection.

Yet not being aware of this scorn of awareness, these short-sighted entities must still act within the evolving, complicated and often stagnant world of mutual realization. To act within such a world can come to require a good conscience for inevitable behaviors likely to occur, to live within an unsolicited environment predicted to follow probable acts of occurrences: especially when required to accept what is unacceptable as understood by any imagined, maturely advanced mind design, rather than a cognizance still within the throes of its evolutionary struggle for dominance.

Within members, these choices are easier to make than for others. The environment and evolutionary traits sometimes merge into a disputed consciousness ripe for such conflicts, inviting to the individual cognizant choices they must surrender to or potentially suffer to death at the hands of the uninformed masses around them. Inevitably, rather than native mindful choices, such an emerging milieu forces these choices made by everyone through highly convoluted formulas of comprehension available from convoluted, invented environments of mind and world.

This transition from a native environment to one mass-producing the insights available naturally in conjunction with environmental symbols, constructed into ideas and made physical in the world around it, complete with a vast range of consideration for the reeling moral condition naturally emergent from it, is inclusive of disproportionate extremes of this condition expressed through the enormous measure of behaviorism of its inhabitants.

This facet of a newly emerging advanced consciousness carries the heavy risk of becoming understated and misconstrued until inevitably it behaves, through massive self-expositions, as the choice of direction of the entire species with scarce exceptions. New members of this species become amalgamated from birth, being born into a world complete with a recognized historical and current condition to which it will be exposed, and come to assume this world, for its prior choices, the sole alternative and condition of whatever truth this newly fashioned collective realization has summed for itself as a culmination.

In this world, the individual, potentially the means of overcoming the species for the future betterment, is given less consideration versus the overpopulated mainstream resisting it: a dangerous pitfall in the evolution of any species utilizing the advancement of its conscious faculty to inspire future evolutionary traits, furthered in coincidence with a convergently inspired bodily form owed an advancing consciousness craving to outwit a complex, demanding environment.

If, within *first* individuals of either species or epoch, a consciousness does become clear and fully balanced in its contextual outlook to overcome the preconceptions of its species, its first wonder should be how significantly the final discovery, itself in dual reflection with the world, contrasts with the means to attain it. The overwhelming feeling of such an inevitable human individual should be mortification at itself and the willful, unthinking evolution and making, through tortuous millennia, of its shared consciousness in a present state with all its past description curled inside – cast as a full reflection of its mind and body's implications conjoined with that history – rather than a different, unfulfilled keepsake of the individual's species during the moment of lighted, inner elucidation and succeeding, solitary enlightenment.

–

The environment shapes and determines behaviorism. With religion or ideologies, a peaceful environment without threatening stresses can create peaceful religions and shared systems of belief or freedoms, just as simply as a violent radical environment often creates entrenched, radicalized systems of belief in response to the same embedded ideas.

The roots of beliefs and ideas are the same as all else pertinent to the survival of an evolved consciousness: its environment.

Beliefs and ideas can counter-instructively convert an original intent or prediction, as they can begin over generations of misuse, usually by those without original belief or thought, to make the future offspring of an entire locality or species come to underappreciate and forget, making of them alternative realities of nature, coerced to ignore what is real in the world, to ignore the disobedient experience of nature, light, and awe as a coinciding sense of oneself and world together.

5. Stratagem Disguised as Rule and Experience

Gameplaying's emergence interprets as an unsurprising, forceful, and fluidly surfacing obligation within the world's evolutionary shifts among diverse species of predation and competitors for niche solutions. To successfully advance in the contest of evolution, it was budding, promising, and selected as a vital, inviting emergence of the fine-tuned and exhaustive improvement of regulated traits as a bourgeoning, identifying validity among the ever-growing expanse of species on a planet.

The designs, parameters and rules of the game are afforded by the forces of the universe previously established before the convergence of life emerged as a new force from a precedingly strewn beginning: from a backdrop of unobserved, energized reality into which life could symmetrically attach and appear with the unavoidable, inexperienced, unrelenting compulsion to play along as an only means of achieving fulfillment of survival's continued betterment, its enhancing focus of attentions and the surfacing of latent talents pronounced through skilled shapes and bodily persistence as a scheme of niche fitness.

To not play along is to not learn, to not extend the unique collections of sensory readiness, predictability, and the willing adjustment necessitated to successfully evolve; or else become consumed and replenished by new, unrealized challengers eager for a chance at entering the immense contest of players competing for niche allotments. These victories bestow the rewards of new successful abilities within a contending species ripe for further play with the confidence of a defined subjugator of personal niche space and as a continued occurrence as a winning player within the scenarios of the game's future.

Through a clever approach to this natural and lethal drama, tallness and prominent physical composition do not obligate victory over more diminutive species and the amassed ranges of strategic potentials contained within its reality unburdened by immensity and shaped by differently perceptive, intended commands of forces.

The exploration of niches is perceived through the outcomes of successful maneuvers organized through established, practiced behaviors and confined genetic routines, the competition of winning evidence against other local and non-local species, organisms and communally created environments playing the same game of survival based on the universal forces they each share as mutual context and gameboard for discovery and invention of stratagem.

The playing fields transformed into distinctive landscapes through predatory movements and attainment of ever newer multidimensional solutions from an originally insensitive environment to one teeming with insights and cognizant endeavors of improvement, continually striving towards ever-greater victories of success and new subsequent awareness for this deliberately acting world of malleable traits, designs and the surfacing of winning behaviors achieved from an originating manipulation of felt unseen forces acted upon to reshape the prior embryonic backdrops.

As an outcome of the playing environments developed through its feats of triumph, gameplaying follows an advancing scale of manipulation, ploy, and primary consequences as far back as the level of cells as descriptive resources of confined ingenuity. It is conceivable to apply to the level of particle, wave, force, and field as a pondering evolution into a workable universe capable of substantiality and inclusive of primal environments skilled at prospecting life through the force of gelling evolutionary transformation.

–

The veracity of winning in the game becomes what realization gets away with by whatever reachable innovations, resources and dimensions afforded as guided paths towards embedded evolutionary transformations; and the acclaimed feat of entering the game as a player by becoming a distinctly competent species within a niche.

The cleverness of skills, the substance and reflection of overawing, antagonistic intentions, verbosity, posturing, and disinterest unavoidably become growingly rewarded victories within the surreal world of humanity where it found such benign attainments useful within the originating world of nature accorded to the simple originalities and native restrictions of reality's subversive contests for niche dominance.

Reciprocally, the unconscious tendency of traits of a prior landscape of reality released through the human conscious interpretation during a progressively innovative future state, such as occurred over the preceding few millennia, can become perverted by greed of rewards, replicating natural successes with dexterous fingers and adventurous imaginings seeking a vindication not otherwise found within nature as an overshadowing, separate, dominating intention: as reachable self-evidence against the validity of all other niche-empowered environments of a planet as a dramatic exhibition of true, conflicting solidarity in consummated, coordinated stratagem.

Higher potentials within subsequent complexities of cognizant awareness and contradictory patterns of reality do confront genuine, realizable and emotive loss as a subsequent defiance against this newfound, unconscionable and evolving willpower to furthering human discipline as a shadowed belief within reality, with each species either opposed or obliging to new symbiotic environments attuned through the sensitivities and willpowers of all species as a competing conjunction, endlessly chasing the transformations of nature's environments alongside the surrendering of old and attaining of new niche markers.

A widely affectionate contempt of nature envisions redemption through the advent of humanity and the furthering of its success as a superior, all-encompassed last solution to a deep, fine, and infinitely detailed universe.

Humanity, as a last solution, perceives itself victory over a planet, over evolution, above reason and cause, and in full, proud belief of its place and allotted role as the obvious end-victor and best answer for the sake of the observable universe.

–

Within the arena of nature, as a restraining rule of the game, success is not all-consuming and permitted a confinement to one niche fulfillment as a dominant, ruling solution for all if victory is to be lasting and not otherwise absent as a force within a welcoming environmental cohesiveness.

The environment must succeed alongside the victors, whether a singularity or multiplicity of continuing attainment. No one species in the natural world dominates in its relationship with all other essential organic niches through unification into one all-consuming niche.

As a result of the silence of spiritual awareness composed of unique sensitivities of reality, this world cannot exist within the scope of a singular collective awareness doomed by isolation without a proper measure of environmental sustenance and its invigorating diversity of associations. One species concocting upon nature a persevering martyrdom and omission in deliberation, or the heedful quest and sought promise of a future model for the perfected, succeeded regulation of nature by its defter hands as an endgame, is a doable solution for a potential species as a short-lived, unsighted triumph utilizing the surfacing power of pride instilled by its original, lasting inexperience.

Humanity assumes a bullying attitude towards an objectified, mocked nature as an amusing, soulless thing beneath a soulful, spirited, meaningful humanity. Humanity assumes the supremacy of its meaning above all others for the sake of its willing reasonableness to make it an obvious truth in a world where previously truth had never been a factor other than change and consequence due to evolutionary victors and losers.

The invention of truth within the universe is for playful endeavors with far-reaching tendrils opportune for stratagem, made for clever minds unencumbered by integrity and the processes embedded in environments through expanding conformations of manifold justness, as awakened beliefs beyond mere truth as a usefulness.

Humanity assumes its truth and its meaning as a symbolized restructuring of nature fit for its use of belief, as the discovering conqueror of truth above all other quieter interpretations of the universe ranging the world in multitudes of variety plentifully everywhere it looked and ventured to walk since its original, proud gamble out of the trees, unwittingly ready to undo an unsuspecting and strangely whispering reality.

As with multitudes of peoples under clever and hard-fought capitulation, nature quietly must adhere and defer to the idea of the perfect fusion of mind and body as fitting the successes of humanity accidentally shaped as the dualistic and sensitive personification of celestial, pure, deity-imaged spirituality within the universe.

Human spirituality, its enlightenment, its solemn arrival as the last solution in full, proud belief of itself: humanity has shaped the universe in its image with the self-assured certitude of an approaching game-winning victory.

Nature does not assume an isolated entity of dominant manifestation within the universe, or at least within an isolated, local solar system remote from further environmental sustenance for its evolutionary branches to reach into the patterned light of the universe.

If there were earlier, primal attempts at overwhelming success at the expense of all other species atop a planet, its long-serving absence of occurrence during any current epoch of its future such as the present *now* of this world should admit its failure to achieve any ambitions of whole, fulfilling dominance against the ambitions of wide-ranging, shared diversity as a better overwhelming solution.

–

Before the arrival of the human species as a surfacing product of nature, wanton greed of removing deviating gameplaying solutions became skillfully contained by nature over eons of predatory millennia through systems unifying the equilibrium of bodily energy and consumption essentials to peaks of coexisting symmetry. These experiences aligned with the familiarity inspired by regulated genetic memories of the solutions through traits to environmental provocations. These solutions produced the constrictions of behaviors and shapes of countless species through the later and merciless games locally played atop a planet for countless generations of predatory change and its innovating improvements into attainment.

Inevitably, success breeds higher realizations, solutions, and truths of reality which consequence through the chance improvement of a conscious cooperative of awareness. This is a hefty risk for nature to assume for the reason of the probable loss of boundary, control, and the promise of an overwhelming thrill of dominance occurring as a confused, willful choice within a newly rationalizing species.

This attentive species, a product of the interconnecting experiences of all prior life, inclusive of the countless species submerged within the planet's encrusted deposits of mistaken or misleading, unsuccessful, or merely progressively strategic pathways of contribution to furthering evolution's gameness and length of field upon which to play.

The innovations of this otherwise illusive species until the point of its upright appearance walking from the waters, mountains and trees, its enthusiasms for improvement through its exhibited, ruthless gameplay of hereditary survival, were ruled by the forces of reality procuring a diversely immersive landscape for it to cling to as a branch holding its finding awareness aloft above an otherwise endless, utter void of awareness, and the game is lost. A void of awareness is much as the effect on this planet and solar system were the universe to suddenly become unappealing to gravity. The prompt correction of gravity's absence would spiritually absolve all physicality of its material suffering into an empty, deprived hollowness for other forces to attempt to reconfigure, evolve or enact upon anew.

Without these prior forces, there would be no choice of dominance among species, no extinction or rediscovery of species, no conception of a potential nonexistence, and no ideas of surrender and transgressive acceptance of a hungered spirituality in need of sating beliefs. Among individuals of a sentient species, there proffers the option of self-exploitation, of perpetrating a con against an unwitting worldly awareness procured through naïve eternities striving towards better prediction of solutions following the rules and scheming play of nature's inspiring, well-rewarded, predatory games.

For this new species, these predictable repercussions spread into all undulations of truths plucked from reality's moveable texture, so that the verity contained within solutions of nature became disguised as motivating powers, as an exalted, inheritably unachievable greatness idolized as beginning, as meaning and persistence, as intention and willful pride of action, and as a playing field ripely fit for diverse conflict.

The inevitable arrival of these historical compromises with reality emanates from the simple successes of traits for betterment and survival among assorted species of life struggling to organize niche conceptions of mind, body, and environment into well-adjusted stratagems. Inevitably, with helpful conditions and innovative, evolutionary energy, there is an ascending appearance of a functioning planet teeming with life-filled victory and success of solutions among an incredible range of deeply textured environmental realities.

Fitness failed to become monopolized by any one or few species, but as accumulations of shared attributes and the occurring manifestations of these diverse accumulations alongside the resultant predatory niche developments shaping the reality shared and composed through incalculable divergent comprehensions. Each identity of species is a distinctive property and exhibition of a realized, successful universe behaving as a stable, current summation of solutions on demonstration for all life to find in proper, awed obligation of benefit.

–

The contributions of traits extracted as mechanisms of gameplay within the natural world become regressed and debauched methods of amusement and self-confirmation within the new surreal world of humanity constructed in its stead. Its growing complexity procures new, differing material tenacities for sensitive bodies, aware minds and the cultured responses produced through believable behaviorisms enhanced by conforming scale and escalating values of attention and deniability beyond the range of a surreal environment's coaxing surrender to further explanation. Its ghostly coexistence is like an endgame won and consigned to persist as a sole residual system to efficiently aid further a sole species bent upon excelling the savoring blisses of its hereditary rush.

Deniability evolves by the loss of attraction to reality, of obscured, unfocused minds acting in conjunction within a limiting, restraining niche environment promoting such values as promising and worthy of attachment for successes otherwise believed absent as survivability.

Localized environments limit and enhance the advance of gameplay solutions due to an incomplete scope of understanding the universe among the manifold singular and conjoined niche comprehensions of its species in relationship with all others, while the localized environments inevitably, broadly overlap as newer solutions become continually sought and imaginatively sown.

Deniability is a quality of the natural world usurped and evolved by the new surreal world. Species cannot act on every occurrence occupying their perceptions as if reality were a feast of preoccupation rather than a game towards utter dominance as a succeeding species among immeasurable species, achieved through niche selection including focus, attention, and confidence in skills.

Species become contained within the realm of comprehended niches allowing the survival of generations worth of hoarded enlightenments. This pre-visioning permits better access to survival as the current representation of a gathered history imprinted within each cell of each member's body and mind convincing it to act appropriately unless otherwise stressed or forced into new areas of betterment or skill in its gameplay tactics.

–

Inevitably, the natural world becomes so distinguished with gameplaying solutions it ceases to be much of a game as niche fulfillment of winning players becomes unwittingly, mutually entrenched within uncountable thickly layered environments. A tolerance of other combining successes prevails within an equalizing peripheral world allotted the eons to advance their gameplaying skills as a coinciding, sacrificial force for betterment. The world becomes a blend of conquerors segregated as species each acting and behaving in confident mutuality through niche communities within varied localized milieus. This concurring feat of abundant temporal experience coats a planet in the comfort of a bright, energized, and wide-arching triumph.

An eventual, overwhelming victory lasts until a consciously unified, willful awareness, generated as a newfound sequence of solutions surfacing from the world of solutions and the promise of immediate and lasting benefit, appears as a better choice of fitness, as a regulator of more hits and wily, unabashed swelling battles of wills against a world of nature brimming and enclosing a planet fully with a vastness of wrought and contented fitness in overabundant sureness.

A world of communal mindfulness generously endowed with the new salivating awareness within humanity assumes limited to no residual relationship with its underlying natural world except as camouflage, and the ingenuity or encouragement of appropriately satisfying, projected human spiritual alertness.

As the victor claiming the right of exclusive spiritual actuality upon the world, humanity gained multilayered access to the afterworld of its conception duly fitted for the human soul. For the first instance within a vast universe, a meaning uttered through human spiritual recognition as seen and found according to the evolving awareness bestowed by consciousness and its unquestionably entangled hunger for self-advantaging beliefs.

This shift in sensitivity made the dispirited old world an observable wonder and mystical inspiration for the betterment of human physicality and chronicled, documented spirituality in the new world.

A conscious species might assume the bewilderment of deniability, easily afforded by the seeming voiceless rapport of reality and the natural world as a selected affinity, for better survival avoiding bodily and mindful dysfunction as species can succumb to as a lost or vacated niche formally comprehended and fitted becomes lost. Instead, as a surviving species to live, celebrate and idolize deniability to overabundance as a claimed right with the attention of self-wonder before a common world.

In nature, deniability is niche acceptance and its motivating invigoration on normalizing behaviors; in the surreal world, deniability is the regression of instinctive values and the discarding with natural basis as a factor in favor of idled, usurped remedies to an unprovable meaning of existence. It is the condoning of this lack of meaning as the foundation for transgressions against an inurned opinion of exempted spirituality in nature, and the phantasm of substituted, prismatic beliefs in its newfound intention for one exclusive, mystically imaginative species.

This deniability can temper on a scale which is unflinching in its restful integrity as fisted solution. Being completely, naturally obtainable, contrived, and malleable, it is a perfect tool with abiding uses. History proves the corruptibility of humanity, that it is not to be fully trusted, especially within the limited confines of any single human being and what might become expected of such corrupt byproduct within the overarching design of reality and the encouraging power it might impose upon masses of individuals. The species could not fast from this unrelenting temptation for rule as a bending source of abundance, so it should not become expected from individuals as the originators, culprits, and victims of jointly satisfying stratagems freshly recurring over a mainstream of historical perspectives.

Coincidentally, the makeup of reality does allow this potential fasting of abusive sensitivity in a shortage of rebellious individuals as a focused experience of identity in behaviors and actions against a crushing juxtaposed world of conformism. These fewer, balanced individuals would become motivated to further or better the future world as lifelong, unannounced self-request. Without the impervious conformity of multitudes, rebellious spirits would find no fit environment or cause to surface as force other than as a merely latent stimulus upon the world of improvement.

The return to nature or nativity as a new wellspring of spirituality and guidance, as a protector of a different justness, as a selfless choice for greater betterment, becomes a strangely aggressive, disobedient pursuit against the healthier interests of a wider humanity and its unquestioning historical contexts of favored beliefs.

–

In the past and current worlds, to learn to participate without the practice of gameplay, without understanding its spirited strategies, urging potentials and intended abject solutions, without learning to skillfully create and find foreseeable innovation ready to utilize as necessitated by the moment, is to be put to disadvantage as a developing inhabitant of humanity or a retreating nonhuman entity or species.

Once, this gameplay only applied in its relationship with the natural world's localized, familiar and inherited environments, but through the socially complexing perspectives of fierce histories and subsequent raised resentments, with technological benefit overtaking the solutions once prescribed to bodily traits, it has become an essential component to survival in the human-made surreal world without having to surrender to nature as an inflexible niche allotment alongside others in unified multiplicity.

Multiplicity within nature as a grander solution conceives as a novice, primordially youthful game traversing the world of survival without abundant accumulation of personal experience or skills beyond the endowment of inheritance from a confined, randomly genetic starting point. The assumption of nature's unawareness and indifference to its fate is ironically undoubted. The niche realizations and conscious solutions substituting traits, rallying new behaviors and aware thinking for humanity inevitably to become all-consuming and wide-ranging in benefitting solutions, become absent further innovative integrity of spirit.

–

In gameplaying situations, the loss of advantage or consistency coming from being unpracticed can be itself a ploy by another for its self-advantage and winning strategy, especially as a handed remnant of the past's convergent power solutions for localities and populations.

In cases of birthright, the disadvantage of this inhibiting style of environmental allowance upon individuals and peoples within an evolving, dynamic, and past-bound reality can leave most at the drawback of starting the game as the victim of inherited strategies poorly devised; or a lack of success merely through historical capitulation to dominant-hungered strategies within the local environment it coincidentally was born into as a new reality tangled and bound to all the old ones as determined.

These environments are equipped and the weakened burdened by the designs of the gameplaying world to previously inspire it, without much chance to overcome the conditions of inherited, perpetrated tactics and its accepted adoptions as chronicled rules and demands; embedded within these same inherited localized worlds of whispered meanings, scaled to the life-imposed values of a new environmental framework, advanced through the relic solutions of contested, historized decisions prevalent in the present world as the spiteful, mocking haunt of a willful force exuded and sustained as an impressively preserved ritual and ceremony within the conditions of abject acceptability.

–

The necessitating nature of knowledge and consciousness is the *idea* of the world versus its *reality*, or the disinformation which naturally occurs from information, of its command and circulation among multitudes of highly susceptible individuals enthusiastic for ready and affordable strategies of new self-appeasement.

The games are perpetuating, with each new generation of individuals needing to rediscover the experiences for themselves anew however many times the similar practice has been repeated by countless others in the past in differing contexts and contrasts of exhibition, so that the textures and subtle nuances of fine-tuned reexperience are exhaustingly explored for higher solutions and benefit.

These reexperiences are dire in intention creating resentful human and nonhuman victims of human deniability as a newly surfaced awareness of the true bounds of contorted consciousness countlessly explored for advantage, and the calculated thresholds of willfulness to assume contagious beliefs uncovered and favorably associated within the networks of human reality.

–

Occurring throughout much of civilized history, this approach to reality becomes a singular attribute of humanity or a convergent yearning of human consciousness in many environments of the same world, as a sculpting of the melded forces acting upon it, permissible through a cognizant aptitude of mutually competing survival as risks, rewards and detriments constrained by the boundaries of scarcely conceived, imposed universal dynamisms.

There exists the marked predatory maneuver playfully regulating the occurrences of resolve upon earlier and later unfounded beliefs and feelings within collaborators. This playful scenario occurs as a direct intention against a marked vulnerability in spiritual confusion and valued satiety of alleged soul once sufficiently belief-enthused within enough historized participants. In this scenario, belief does not secure as ownership but becomes given as a provision and commodity, and the idea of its necessity for genuine sustenance and unassailing devotion ripe within the physical world of a surreal-minded humanity.

This devotion to spiritual satiety often leads to the instilled habit of comforted sameness with fellow partakers, while accepting the confounding sense that delinquency of conduct and endgame strategies are tolerable if it helps the self's group as a primary fitting success.

This duplicitous thinking is not moral thinking since any natural conception of morality is shaped by the expected multiplicity captured within a species exacerbated by an advancing knowledge of conscious responsiveness and enhancing focus, with its processes of emotively thoughtful responses mimicking the preconceived methods of solution reflected in evolution as diversity and the niche innovations within manifold natures of concentrated awareness.

—

As a species, humankind has learned, as a means of better solutions, to disguise itself from nature, stepping out of its skin with an intent on shedding evolution from its future as a further, peaking realization. A species of individuals is mutually disguised in the full spectrum of observation before nature and its celestial eternity, unafraid of admitting what cannot be recognized, unafraid of its limitless inexperience, and its hollow enlightenment for the horror of spiritual interpretation within a freshly correlating mind; instead, finding it as a confidence beneath the disguise, deriding the natural world and the cosmos for its nakedness.

These disguises of character create a surreal world living atop the old real world of nature and existence as a discordance and convenient, communal denial.

The disguises are costumes of mind and behaviorisms of personalities in a firmly established environment. They are the mimicking of fitness for the mind to endure its new world no different than the disguises of mind, skins, furs, clothes, shells and manipulated light found in nature permitting the continued surviving longevity of its untiring inundation of overflowing species for so long.

—

Throughout human history, there is the progressive beautification of self, the virtue of self, the power and will of an identifying personality mutually among the array of other seekers of the same identification. It is naturally peculiar to assume such will and design upon and for no grander motive than oneself among multitudes of others doing the same as an endgame of the universe.

Attribution belongs to a natural lack within reality, its absence of certitude in direction or practical guidance otherwise without adhering to the interpretations of an intuitively silent nature and cosmos observably hollow of any other plain answer or unifying meaning, contrasted to the plentiful answerabilities of surreality. This occurs among a backdrop in which inevitably we no longer look to nature or the sky as finding source of promotional awareness through survival.

Humanity, from an origin, became haunted by allegations of spirituality and the descriptions of chronological events within a world devalued by the unceasing surfacing of personalizing identity in a vaguely correlated reality, as humanity attempted the lengthy process of overlooking and forgiving the natural world its usual disobediences.

The world of knowledge being equal to the world of spirit in value and urgency, it began from a foundation of unawareness, forcing the obligation of initial and compounding predictions, associations, symbols, and meanings discoverable as usable solutions for the manifestation of humanity's innovative ideas of the world. This fresh world originates as an inexperienced, absent spirit of the natural world attempting to survive the new associations of mind and body compelling it, just like cells regulating and producing proteins for active improvements, to find and construct better, materially substantive solutions.

Individuals prove the drive of individuality among a horde of other similar accumulations of intentions and survivals, concepts, and beliefs about identity of actions through the echoing patterns of generational reality.

Life becomes for humankind a stratagem for the establishment of sequential rules and visions dictating the arguments for new streamlined experiences ready for concentrations of localized offspring to aspire or despair. Alleged redeemed solutions as something snatched from reality, as conscious workable disclosure from past inexperience, the transformation of nature and its impressions ensues through wills ravaged by the many processes meant to contain disinformation by fostering and recording it as ownership, as the striven measures for the supervision of spiritual beliefs about individuality and otherworldly losses and rewards as the means to further intimidating success in this world: as an old persuasive force or stratagem successfully lingering within the world of human stipulation, as a meaningfully projected determination and the regulating measures of synchronized guidance as benefit.

–

As a facet of humankind's overexerted advancement as a willpower and individuality upon the world, acute individuals will come to perceive not only the present but the future as an intention, tendency, and coincidence, finding it to be of greater importance than the urges of a faithless, unreliable present state.

Through force of subjugation in advancing absolute knowledge and spiritual values, individuals will leave personalized indications, ideas, contrary perceptive tendencies and conceptions of instruction or validation for the future to preferably grasp what a wavering present state of humankind as a historical era could not except within the imaginings of exclusive minds differently cognizant enough to predict and attempt to control it. These individuals admit the futility of any lasting revolutionary conversion during the presented world of visibly unified inexperience claimed as a determined, unceasing present-day reality during their era.

Within this scope of human descendancy of spiritual and contextual inheritance, there are acts of great purity and insight, and acts of great spite and retribution to be handed to recurring generations.

The future should come to inevitably, passionately sympathize with confident individuals of the past with admiration and recognition at how they at least attempted to benefit the future through envisioned communication, risky actions and zealous illustrations; and the many with alternate sensitivities who had to endure the worst atrocities of the continuous present state often dominated by a forceful will to punishment against the natural world of nature and clear-sighted attunement with the advantaged partiality of humanity.

Instead, humankind's spiritual nature clearly arises in the situation of its gleaning image as iconic motivation furthering the steadfast paths of intimidation and verdict found inherently embedded inside the making of current human and nonhuman reality.

–

A species can unwittingly set itself along a path of eventual defeat by nature and its methods of longsuffering regrowth. Through attempts to banish empathy and inspiration with directed focus on an obscure mainstream end by whatever means, forced to admit to submission to the evidence of abstract leadership as the redeeming talents of personality, given a choice of unrealistic options, people will usually find the one which helplessly validates them utmost in the moment. Justification of immoral behaviors begins within the nature of personality and spreads through the reflections perceived by others as commonness.

To survey and study human history is to also realize the future's past potentiality in the current world with an improved, unbiased predictability and a fresh dissociating sense of loyalty to that version of impending history. Understanding in others the ploy of manipulation, that to pity misfortune in another will shape and assume that pity within them as an identity of subjugation or revolt. It is the same with all else shaped from the peripheral force of experiences. As it occurs in nature with evolution simulates to enhanced extent in humans beyond the origin of fitting intentions and rigorous solutions.

It argues that morality has its place within laws of nature, or epics of time, through its manifold descriptions of brutality, but nature and epics of time evolve the same as all else.

The castigation of a puritanical worldly acceptability in the mystical phantasms of true belief should inevitably emerge through relative safety of societal ease, through the growth of understanding of its validity and misuses, through communal and individual self-passions and an unbending acceptance of mutuality within a discreet universe of human and nonhuman awareness.

—

There is a discovered conception of spiritual meaning fitted to human consciousness from the natural world, an emergence of universal meaning endowed within the personas of humanity's participants.

Humankind exists as sole ownership and in assertive belief of its entitlement of spirituality, of soul, of awareness and passion, of emotive thought, raised at the expense of the reigning balance of a natural world, a natural universe, of limitless species established as a prior icon intended to be swiftly uprooted with elevating victory to humankind as a pride, to rid and adulterate its hard-fought and raging attainment in poised competition, just and worthy of expelling such dissenting spirit into submission and the oblivion of unawareness with an ease not applicable to integrity.

Humankind can exist to take on reality in all its facets as the wellspring of otherworldly trial, as the redeeming, arriving life-force and proper endowment for the future as the fulfilled, instinctive realization of a successful endgame strategy, a believable triumph over the physical and the spiritual worlds of reality prescribed, for lack of better alternative, through the interpreted representations of the natural world it reciprocally demeans as unintendedly antiquated, unfitting and randomly accidental in its tactical, strategic solutions, communications, emotive awareness, and spiritual values, as pawn atop the lush landscape of ranging mountains, forests and waters making up the vast closed fringes of its gameplaying fields.

6. Consequentially Derived Choices

In a world premeditated by past choices, there are dynamic aftermaths to individuals and the human species generationally acting, fittingly behaving, and scarcely knowing the reality they evolved to be part of. Without broad detection, the world of humankind repeatedly loiters while furthering its own predictions, without imaginative seekers of finer, better truths resting blatantly dormant, and its participants stop admitting the underlying natural world as magnitude in its connotations as a living, breathing, responsive awareness. This undervalue of substantiality mostly occurs in favor of simply accepting the nearest local facsimile of iconic belief according to a benign allotment by birth.

This agreement occurs coincidentally or convergently among a diverse range of individuals with intended members grouped by historically distanced localities of both time and space. These originating distances limit devastations potentially and probably caused by minimizing contacts with outside and competing human species, permitting familiarity to appease the shyness of members as settled recompense through its fitful mollification of populations of undeterred spirits living as the coordinated and genius wonderment of nature's exhibitionism.

With humanity's believable ideas of the world, this randomness, chance allotment, lottery, or straw picking of spiritual and moral value worthy of honored actions appears to be lacking as a reliable target. It is worth more as a mutual, common value marked by possessiveness and self-justification for the sake of locality and its deep-rooted, culturally social interpretations. It exists in awe of an otherwise absent or unpresentable unifying answer.

In these cases, coming from the same world and limited environment of the universe, these differences and resulting conflicts of distrust are symbolic in nature and through genetic grounding cultured accorded to prior achievements and pragmatic conceptions of local successes; not as an only path or means to success, but the one previously selected and retained due the conditions and events of the surrounding, mysteriously living and nonliving reality it evolved alongside.

These systems of belief inevitably become potential struggles of conciliation as the old means of holding value in spiritual management is overwhelmed and reinvented by the natural improvement of an evolving unification of multifaceted beliefs, the ensuing insights into the past which become revealed through higher understandings, and the reduction of the human fear of others accorded the inherent risks of trying to survive a stacked reality manifested as a natural world veiling another, inner, reclusive world of bestowing attention.

Sufficiently advanced, the human-tended world begins to become overcome by the sharing of spiritual and physical traits, with beliefs stored as psychological solutions through generations of choice-fueled moments during any present state of the evolved past and repeated until the present state of a perplexed world.

The amassing and detailing of chosen solutions of belief among groups can disclose to be similar if not chronologically coordinated through coinciding isolations and contacts of diverse localities of history in varied stages of conscious willingness and acceptance of reality. Sharing the same universe, world and environment of niche-constructed life, the similarities of structured beliefs will be culturally and artistically dissimilar, though the roots constrain by the same urges of the same universal forces and spectacle of prismatic reality.

Humankind was presented, during any past moments and whatever the eventual outcomes to the present world, with the same or closely consistent environments of life, with the ingredients and materials to reshape into the implements from which to create, identify, decide and grow as a unified version of evolutionary culmination identical in valid attainment and indebtment to the natural world and the unified mutuality of nonhuman species acting in coincidence for so long adjacently together.

—

This possessiveness given to beliefs and the restrictions of choices derived from prior chronological punishments is a lingering remnant of a closed, isolated world preserving its boundaries of assorted claimed ownerships. As these boundaries of distance temporally dissolve, they become defensively reinforced through overlapping contacts into what is finally disclosed as a vast allotment of localized, environmentally and historically founded belief systems and humankind's reclusive, entitled revelation of an enduring spectrum of multifaceted mysticism, placing itself as a celestial epitome or center presiding in a present state of the world in every occurrence and all experiences as a gifted rapport. The notion of humanity spotted as a center of a universe stays unremoved from antiquity within a current present reality due to its soothing successes upon the human awareness for its soulfulness as a descriptive and flattering mirrored self-image.

This unification could be an act of pride, gift, and vision, or one of retreat in defense against an unavoidably approaching, closing future.

Given the undeniable choice in most cases of inevitable amalgamation, mystical isolationism and minority entitlement within clans and populations reaches a peak of success as a better solution. As with evolution, a change in trait or skill is never immediate within a species but takes exploitable generations to become situated as a newly surfaced property of its proceeding social reality.

A rational species, with a true measure of rationality, would view this inevitability as a natural movement; instead, humanity perseveres by safeguarding outdated systems of beliefs as if a survivability depended upon it – which it often has as mutual self-fulfillment – but it is the current habitual labyrinth of past choices of survival as a best and only solution not to be reneged. The distrusting, unknown risk of choice and the permanence of choices within other individuals and groups demanded enthusiasm for the past as the unswerving answer to the present.

Conformity to belief in the new evolution of ideas becomes resistance to the future for the betterment of the misrepresented present: with only the past misguidedly thought to relinquish. This belief of the world persists until its awkwardness finally reveals itself as unsustainable within the realizations of a far-distant coming.

Through new thinking imposed upon the participants by the observable and chilling shifts of its environments, as an inevitable consequence of the old, ill-conceived choices of mindfulness imposed upon the world for so long, the environment begins its stabilizing response in scale as a karmic, reckoning force ready to make its sardonic mark upon an obliging humankind as a fatefully twisting resurface of equilibrium.

–

Human history can hypothesize as the admittance of its original denial and the surface of humankind's pressing starting illiteracy manifested upon the symbolic landscape of the preceding, heeded universe. Consequences of denial, on a bodily and spiritual level, over layering generations atop generations of mostly choiceless identity within manifold experiential outcomes, with captured risks absolved through an initiated, conditioned psyche imprisoned by the parameters of its successes; and the quiet regression of its sensitivities through the shift of its skills in favor and inarguably capitulated to the sole domesticated mystification within humanity's rightfully asserted mortality and reclaimed otherworldly adventures.

Nature is forgiven its prior eminence within the human being, giving it the strength and vigor, the soundness of emotive certitude and proper deliberation, the behavioral routines of belief, and the conviction of solution within itself to stand alone as the intrinsic value of reality, as the fit arrival through the shedding of evolutionary consequences, and proudly admit to its tamed spiritual being as the gifted spiritual, emotive and mindful reality within its scope of the universe.

The consequences of choiceless spirituality, absolved of doubt or further embellishment, into a world of pure confidence of belief in what has been handed down by exhausted generations of temporal triumphs and prides. The current present state restrains its spirituality by the incomprehensible array of unprovable, symbolic beliefs of one overly reinterpreted environment acting in conflict through the confidence of localized elucidations being the emblems redeeming humankind's ongoing choices.

This current, presently unkempt state of the world stretching back the spans of millennia, should be conceivable proof of the failure of spiritual value rather than any measure of furthering success. This should demand an evolutionary shift in overcoming by embracing the antiquated origins of spirituality and its embedded imprint as a regressive outcome for promoting the physical manifestation of humanity as the inevitable, singular self-image of a world founded on disobedient thoughts and unetched beliefs claimed as truth.

In this thinking, there is no regard for the absences and voids of nature in store as sustenance for the emotive thinking, the patterns of meaning, and the symbols of belief for future generations condemned to nonexperience of nature's confidently enlightening essence. The condition of humanity's future becomes shaped by a world absent of natural context, of natural association, of shortened prose and creative backdrop for art, and without the natural world's insightful cues and hints furthering scientific discipline.

Instead, an impending barren creation is proffered and handed over as the chosen world like an ornately framed vacant landscape, inherited through the consequences of prior generational collusion of wild, vindicated inaction and the degradation of further incoming meaning or natural whispering for the future of probable solutions due to a willful efficiency and consuming fulfillment of the past finally absent of other than human spiritual awareness in an idyllic, lastly realized present state.

We reason for our growing social evolution through methods mimicking the evolutionary struggles of nature. Both construct from the same stuff of reality, whether genetically inherited or relatedly sought and experienced.

Choices force a retainment of consequences, but consequences also force a retainment of eventual enforced choices. Consequences restrict and coerce future choices of entire populations of individuals with genetic reverberations gauging its outcoming realizations, regulating the intricacy of successes in advancing a future niche certainty for offspring to emerge into as a concreted world of human dominance atop a diminished world freed of natural experience, including a new humankind emotively attached to utter spiritual nonexistence as an inheritable, mystical enlightenment.

The risk and reward stratagem of evolution, niche compositions, degrees of conscious use, and the consequences of excessive choice upon a species of nature can be its undoing or its success in a skillful universe. Realizable choices limit and unemotionally bound to the consequences of all prior choices and the resulting behaviors they reap no differently than an identical genetic brand and held memory within each cell of a body, and the behaviors set by this vigorous confinement of alternatives pressed against its defining fate.

—

It is a rebellious and unfounded envisioning of a reality where the individual walks a world and existence absent of justifying, inborn consequences in poised unison with other individuals, without the surfaced rewards of fraudulent personalities, transgressive leaders and privileges as representations, where the highest rewards of humanity are profitably accumulated by the basest, dominating and intimidating talents rather than its many more substantive aptitudes and remoter experiences of reciprocal human and nonhuman distinction as evident belief worthy of undergoing collective action and fitting inaction, and the higher behavioral adjustments required of unbiased ambitions.

These isolated individuals readily imagine existing in a far distant past, capable of extreme feats of perceptual connectedness as precursor to an acclaimed spiritual possessiveness to come later. They were emotionally attached to the natural world through a nonhuman bond with a textured, lushly pliable and survivable world, possessed of the memories at the newfound wonder of unprecedented events and the blossomed spectacle of reality to observe as isolated spiritual individuals consensually standing fully upright, heads cast upward eying the reflecting universe without alternate purpose or fitting design upon it beyond the moment of sensitive, absorbing observation.

An existence acts in mutual isolation and attachment alongside its living manifestations. This removal of nature's freedom gives for collective benefit, not the advantage of those simply willing and enabled by the prior forces raising its reality to do violation against fathomable weaknesses in others as an intimidating service to its immeasurable self-fulfillment.

Importantly, what a portion of these other violated individuals will not commit, out of consequence for reasons of integrity or the furthering of spiritual success and example, perpetuates the erosion of the best markings of humanity into a weakness and subjugation absent its required forcefulness due to a necessitated mindful appeasement of an old, sympathetic conscience of nature.

Remnants of these old feelings will endure and haunt sensitively prone individuals. During any one time, there can be convinced individuals to see more accurately their present and predict an inevitably mounting future to greater depth of vision.

During any past eras, rare persons will come to detectably view the impotence of a current knowledge in transition as limited in its ability to attain an answer with any immediacy within any of the countless, enamored present conditions of reality. They foresee a future in which the solution might become grasped. They must live and die without reaching a convincing meaning. Instead, forced to exist as a waypoint along a path towards impending meaning for others to inevitably claim. Only in the dignity of acceptance can they reach a meaning found as an absence during their lifetime epochs.

And to view all around the mocking illusions of shadowed meanings in unavoidable, prismatic fruition, each demanding to prove itself the spiritual actuality of celestial perpetuity.

Without certainty, the sole redemption becomes the contrived and imaginable, or succumbing to the localized guesswork of nearest faith and its belief in a sensitivity of inner reality as the path towards any enlightenment of continuance beyond death.

This alluring threat of redemption becomes of utmost importance to the newfound consciousness of humanity it existed without during the millennia of growth alongside nature's nonhumanity as a mere attribute of something phenomenal in its occurrence. This amazement happened within a universe only managing to divulge and excrete the methods of substantiality and the appearance of life atop a spinning globe within a relative infinity of observable reality.

—

There are consequences of human actions, choices of morality, conflicts of belief, and exhaustive opportune reasons for abuse perpetrated upon nonhumanity and the rest of humanity through its senses, beliefs, emotive thought, and survivability as continued evolutionary solutions.

As our knowledge and advancement grows, so too does nonhumanity evolve alongside us in findable spaces of attention, perceptions, and sensitivities when it comes to an increasingly fractured natural world it must often navigate without the full comforts of its niche selection and former accumulated successes providing it with full stature and confidence. As environments shift, species become more victimized by lost niches, making of nature a collection of misshapen, unfit scavengers for the last bits of attentive reality before the fading awareness is gone due the overwhelming, dominant successes of a chokingly original, unidentifiable, insensate environment.

The human world becomes compounded by the risks of external threat and the unknown the same for all species within the natural world, and by the same standards of bravery or cowardice while evolving before it.

Recognizing our capabilities as a species, as measurements of accidental self-realization, without being perceptive to our capacities as a psychological invention of nature or the heedless reshaping of reality according to an image, as solitary remaining deliberation of a fate, our good fortune is lying behind and ahead in potential and range as an evolving exponential threat to the natural world and the achievement of its overwhelming prior feats becoming overpoweringly lost, or at best unwittingly shifted to a distant, more hospitable, temporal future to convergently reoccur.

—

The deniability of generations not collectively standing up against human history as an atrocious, *diagnosed* place, yet persistent in force in the present state of the world the same as it always has been since the birth of our original ignorance began seeking the myriad truths hidden inside reality waiting to become challenged.

Learning to avoid admittance, through strategy of acceptance, of the consequences of actions and inactions, so that deeds become willfully done, so that an unfettered humanity can be less informed in its fitful becoming, unburdened by the implications of a lacking integrity as a gifted future.

A world where becoming an identity is a willful choice of being a singular human being, rather than the many forceful alternatives coerced by the circumstances of appalling outside constraints, due to other wills less restrained in choice of actions and behaviors, constrained now not by nature but by the pre-established surreal world of humanity as preeminent value.

To make a wasted misuse of human history contained as the embodiment of current disciplines of human achievement wrought through eons of delightful, restricting punishments for the betterment of stemming the emotively freeing world of ideas from enthusiastically billowing.

No species, with an aim of achieving enlightenment or anything resembling a goal or purpose, ever acts alone but in conjunction with all species within the present and its evolutionary history as a planet within a universe starved and longing for the recognizing discovery of meaning, rather than the exclusive sightlessness of its continuing absence.

Though currently eminent in its self-status, humanity is not a monopoly of thought, feeling, or consciousness. We are simply a present time and place along the advancing lineage of this one planet. We are unsolicited other than through the accidental choice of our bodily shape and mental design, one meant inevitably to appear.

We are much less chosen by an omnipotence than we are designed and naturally evolved by the forces of the universe into a bodily shape, mind and environment, as a chance at singular enlightenment, where the true individual becomes a dynamic enthusiasm spotted in immeasurable space meant to hide it, alongside and the same as all other species in unbiased commonality, counted among the countless attempts at experience, the multiplicity of budding illumination and inklings of lighted, emotive amazement as living, grounded, and placed beings within a boundless cosmos fortuitously rooted in evolutionary actions of reality.

7. The Encouraging Deniability of Belief

The confidence of spiritual redemption in this world and an otherworld is the certainty of our opinions in action and physically crafted through the gifted elucidation of the human edifice, a world planned by human reworking, raised by the breaking toil of its demanded reality, with the proof of its solutions as iconic ranks and presumptive ideas as have been judged worthy of approved representations through the astuteness of humankind's vivid reason and especially emotive sense of awareness.

There can only be redemption following a path of humankind's ideas and sensitivity to ideas, since the connotation arises solely through the establishment of human supremacy navigating the diverging paths of truths, occurring above nonhumanity's tacit subjugation, as it practiced and stated to be the utmost vindication of the world and its celestial associations. Nature is helpless against the rousing voice of humankind.

No monopolized redemption of spiritual apparatus before existed in the world until the upright postured human being with its freshly shaped cranium permitted the revelation of its spirituality to be the awaited context of worldly reality since the beginning of the universe; as it became coincidently alleged, crowed on by its bulky mystical endowment, and thereby divulged with outrageous realization over the most recent millennia as humankind's burden to seek to fulfill.

Original scatterings of humankind carried it with them from the forests, waters, and the mountains: the feeling of one whispering spirituality. Wherever it became an occurrence, humanity proved the same one, redefined, reimagined idea of unworldliness founded on its original emotive instinct for nature distortedly refocused upon its own conscious self-image through advancing its imaginative investigations of a strange new world.

Inevitably, it gained by revelation a visualized self-origin in coincidence with time and the universe given the confusing characteristic of unpracticed, infinite, and newfound awareness of mind and emotive devotions of storied accounts.

With ample generational space within which to act, permitting it to focus and configure similar essential spiritual and humanized embodiments of a newly founded rightful place and consequence of the universe miraculously unveiled, this mysticism of mind struggled with an abrupt boundlessness of everything concerning mind and reality tempting its inclinations and believability. The only limits of this abrupt announcement of infinity are the restraints of its universal forces and the evolvement of its local environments anchoring it to the finite as a fixture of moveable appearance.

–

Something, such as a simple symbol, which has meaning to one individual or group is likewise meaningless to another to the point of being worth the usefulness of posturing, intimidation and violence in matters of safeguarding the spiritual blood of life, even when the intended spirit is merely two portrayals of an equivalent value of meaning, of the same exceptional, peculiar interpretation of a physical and mystical environment.

The meaning of the world's understanding is by interpretation throughout history, with dissimilar cultural and murmured articulations of the same emphasis of reality worthy of impassioned rallies, fighting for and against the impassioned rallies of others as a nurtured instinctual imitation of nature's superficiality of physical fitness in choiceless engagement due the shaped limitations of bodily and inventive challenges.

–

Transcending this life as factual: there cannot be one truly lasting, dominant, favorite species. The path of one is the incorrect path of another, with equal validity like a particle's identified wave and pointed reality as multiplicity.

Embedded within humankind's ideas and emotional interpretations, as with nature as fitted niches within overall diverse environments, there must be a communal of philosophies like traits and sensual communications woven as a bionetwork by its species and inhabitants in conjunction and as accidental coincidence of willful unification. The ideas must organize into a cohesive gathering of trait-like functions and sensitivities for the entirety of an environment atop a planet meant to evolve a means of genuine survival ranging a temporal moment and a singular placement within that moment.

Nature is devised of many environments filled by species through established niches acting as a unified force of betterment, and true individuality restrained to the wholeness of species as an entirety without superiority of a singular environment, or the overawing dominance of a primary species of nature considered sufficiently attentive, intentional and willful in its penetrating designs.

—

Redemption is not surrender, acceptable avoidance or inaction as an inner premeditated temperament embodied as resentful or embraced sightedness of the identified weaknesses of a vast humanity and nonhumanity vulnerable to its ploy. This detachment from the rest of reality occurs without deliberation of contrary sensitivities and awareness due to the condition of biased localized *surrealists* preying upon environments as dominating predators.

Any true concept of redemption in individual and species is not forgetting the motives and locations of mountains, trees, waters, and stars as valued and intrinsic to the whole, as consequence and source of any fitting spirituality and alleged chance at either enlightenment or forgiveness.

Redemption of individuals and species does not have to be an attribute of reality. The universe continues with or without the encouragement of human or nonhuman evolution and involvement with it.

Redemption involves admitting the circumstance of the compelling unknown, of an impending, approaching future valued with attention as a reoccurring equilibrium. This lacking advanced identification of our spiritual and physical manifestations clumsily cast in inanimate stone, iron and wood is our collectively mimicked, approving cheer for humankind.

This occurs alongside a life-force and universe capable of congealing an original sterile planet into an advancing thickness of proven evolvement and brimming environments justified by success within each cell of each currently acting species among multitudes in improvement over a duly diminished past.

Few individuals truly find and indulge the species in such feelings of redemption, since few individuals can conceive or possess the power of prediction, imagination, and the confidence to view the species and world as a chronological wholeness through innumerable times and spatial locations exhausted by the cycled nature of life's shared existence.

As history and knowledge progressively uncover, what becomes revealed is humanity alone in an incomprehensible universe far beyond its understanding in scope throughout its succinct history. History proves humankind as a verging threshold of something imaginably higher within itself while perpetually conflicted by the manifold divine ideas of this interpreted imagination as an endless conflict over an originating spirituality taken from nature as a derivation; then compounded through scale of unlimited inexperience, collected metaphors and amassed events of dissociated histories.

Similar to how evolution can permit branching from an originating species into new similar and eventually dissimilar species afforded by prevalent conditions for new niche realizations, human spirituality has spread through wide and collapsing distances, its roots of origin from the same highlighted world, with its fluctuating environments of shapeshifted mysticisms, and abruptly exposed as a plaything to new localized versions of an adroit consciousness with pre-embedded tendencies of instinctive superstitions.

–

Inevitably, with closing expanses of knowledge and growing successes of strategy, a new world will have to progressively admit the old world was never real due to its misrepresentations of physical and spiritual natures, yet its meaning remains within present reality as a hindering deniability falsely existing in disbelief of its own confusing reality rather than rightfully nostalgic and listening.

Eventually, it would become recognized and accepted through the growth of admittance to telling physical and spiritual natures to be the propagation of ideas, myths, human-made facsimiles of intentionally misunderstood human-specific meaning, taken in symbolic mockery from an all-encompassing natural world for the purpose of retreating from an unadmitted origin.

Instead, reimagining a new world where all the base spiritual motivations for human action are absent, where human spirituality is not the epitome of realizations within an endless universe, and all choices of the past and present abruptly become primarily empty of value other than human dominance of an environmental reality for its befitting purposes.

The honest measure of deniability and what it can achieve as behaviorisms and beliefs finds its countenance through the advent and incomplete surfacing of a certainty in human supremacy and this resource of spirit as motivating justification for any alleged consequences of stratagem and a relinquished demand for the final challenge of an endgame win either in this world or the locally accepted rendering of another world.

Beliefs are insubstantial and carry no true weight within the believers they each solely inhabit since the unprovable beliefs defy physicality, supporting the provision of meeker explanations and an appeasing attention to toil.

True belief in action according to true value surfaces within reality as something unnatural, anarchistic, and selfishly recovered through self-sought resistance to a normalized world of meeker explanations.

Deniability becomes defeatism juxtaposed as success in the *surreal* world, with accompanying surrender of mind and the renunciation of the world for the absurd sake of an approved, otherwise unidentified, unchallenged self-meaning in an otherwise cold, uncentered universe.

–

Baser sentiments are a regressed revilement against truer natures of being, or the distortions of humankind's idea of its spirituality started from a premeditated, far distant beginning, perpetuated through ensuing generations of history into a conviction.

Since the multitudes throughout history must accept any betrayal of nature whether against human or nonhuman, including delicately shocking actions concocted against fellow members of species, they cannot accept a gentler, contrary sentiment to emotively consider without admitting themselves possessed of the unfocused conscience of a perpetrator. Individuals within this accumulation of people and long ages come to think of themselves as a truer nature, as the agents of the physical and spiritual fireworks of this existence, adorning in this honor before the cosmos as a justification for means to an end action.

In earlier epochs, life was shorter in which to seek survival and occasional enlightenment as self-attempt at a wholeness of experience. Enlightenment as redemption envisioned as denial of experiential countenance, as inner self-reflection contained as a constancy uninhibited by the reception of opinions from the outside world.

The birth and origin of the rare individual seeking spiritual denial of all prior attempts to self-disclosure by the species become healthily contrasted against the skilled, popular creeds of humankind's idolized spirituality as advanced model.

–

Human beliefs, like all else human beyond instinctive physiology, as a recent phenomenon within the world or, through an extreme of imaginative potential, the entire cosmos, is not fully instinctual but cultured beyond any genetic resistances to contexts of environments and histories of differentiated experiences consuming a prior confidence in nature while instilling its own brand of sureness.

There are varied, manifold states of aware, wavelike realities based on species, the forces and symmetries of existence, niche-consumed localities and embedded, structural environments, histories and directions of evolutionary choice among a preestablished library of innovative and convergent solutions to reality's dilemmas contained as an impression of a world and its envisioned universe atop a compulsively whirling bubble ensnarled by and as the symbol of one sought existence.

As with a writer creating a book of fiction, word by word, line by line, and chapter by chapter consumed with emotive ideas, the experiential familiarity creating the words is the latency of existence, there waiting to be shaped, limited in potentiality by disinformed beliefs as difficult to disentangle as reality itself in all its immense and spirited pride of composed associations defining the solutions of celebrated species like ancient tomes of great, mastered art.

–

By a peculiar twist of belief agreed from past ages, humankind as the spiritual epitome of the universe transcends in death rather than in life. Death becomes the alleged source of measurement of a feared justice.

No one species such as human beings, being an embryonic, willful origin of nature, could transcend the universe, the world, nature, or enlightenment into deathful permanence founded on self-purpose or self-glorification without true merit of disbelief and strategy of denial. Yet, as an appeasement of the senses and imaginations, it becomes the roots of environmental becoming within the human realm of physical transcendence into spiritual being as the spoken beginning of honored networks of belief.

It would be more pressing and fitting to see the act of existing within a complicated reality as the enlightenment to shadow and aspire. By the design of the universe, the value surfacing from destruction usually must precede the pursuit in intended imitation, otherwise there is no original conception to follow and seek like the lure of a tempted soul plainly imprisoned within a transgressed body awaiting the awakening of mortality as choice for reprieve into wholeness.

Within the realms of complex potentialities and forceful evolutionary compulsions required of an especially sensitive conscious spirituality, this belief can lead and imperil an entire world of ideas as it pursues, as overwhelming stratagem, an unsuspecting future confounded by true, surreal beliefs as mentor in the detection of intended actions and the designed, mystical asylums for appeasing the long intimidating anticipation of a welcoming afterworld.

This method of stating the world enables and empowers the non-believing, non-spiritual, non-empathic potentials of a talented humankind consequential to an aged, embedded design found to act supreme and as a fitting success of posturing, embattled actors upon the world and representation of humanity and nonhumanity as its answering, scripted stage drama before the celestial audience.

–

Morality as normalcy cannot take true, natural spirituality out of the equation into purer humanity, any more than it can engage in its associations as solely idyllic human endeavors of heroism and confident devotion without the success-woven conviction for nature's nonhuman world existing alongside.

The proportionalities of human progress are built on the endurance of now-labelled, categorized and normalized disorders as attentive connotations of truth, of spiritual-minded individuals forced to suffer and quietly divulge a contradictory perception of normalcy, as entertaining human mysticisms of spiritual instigation, and strange practices of manifold ceremonial beliefs in a common, current state during any of the apportioned past's many located states of innovative devotions. These shifting spiritualities attempt to persevere on a level for the betterment of others, especially the future as an only endowment of self-assessment during an individual lifetime without motivating affection for its meandering afterworld.

The only solution is motivation in belief by reality and its prismatically announced, communicating description.

Instead, the alternative is a caricature of human invention trying to highjack that same naturalness of belief in existence with compulsive, indulgent aims. A mind and body must become attuned in empathy with reality as its reflection, not through glorification or the mimicking impulse of sacrifice and fear, but through deliberate vision and instinct of individuals among collectives and a wonderment at its endorsed celestial and pulsing actuality.

–

Human history has been excessively restrained by the justifications for choices and behaviors left for the future to divulge, an abusive consequence of the polluted idea that an unlived, unfelt eventual future ends will justify the means of present actions as a lopsided equating of a symmetrical reality, as if such consequences were predetermined, ordained and not merely imagined in fullness by and for the involved participants of an astounding, divinely rapt species.

Generally, no one has the capability to solely predict even the near, local future by current actions with high accuracy or depth of honest confidence, becoming further encouraged to ruin the longer the attempted prediction seeks beyond the immediacy of a fleetingly stabilized, replenished moment.

–

The disinformation of humanity goes back to as far as the beginning of its rapport as a malleable consciousness, beyond the first tools embedding images on stone. It starts at an origin extending outward to the present state of the world. All divergent successes become ignored as unbefitting unless disclosed as usefulness as an imitating ploy, forced and constrained by the environment's composition and a past filled with funneling, believable selections.

The impulses motivating conscious minds are multifaceted and range the far recesses of potentials in its seeking after itself as a proper meaning, enlightenment, niche or divine origin; but the atrocious extortion is the same in all debasement, in each surrendering self-belief, in the resentful futility of acting as an individual against stacked odds of finding genuine, confirmed accountability of spiritual rewards within the vast range of physical validations.

The spirit of reality is the invention of a limited opinion of reality exposed through human doubts, inhibitions, fears, and uncertainties exemplified in the surrounding environment its consciousness awoke within as an originally proud, expectant confidant of nonhuman nature.

This proud, original human being became fitted as a deity for the natural world no differently than other species in their niche reality. As an accumulation of belief in the unification of all species, and the certainty of its emotive insights, its beginning sightedness instead oddly redirected upon a mirroring reflection of its spiritualized personification as sole embodiment of the universe and the embedded epitome of perfected, long-sought answer.

8. Humanity was Once Nonhumanity

Nonhumanity and humanity are mutual and opposing domain of the same motivation for improvement or betterment. As currently this awareness minimally understands due to human bias residually remaining from past generational values, humans were once nonhuman for vast, unhurried millennia, and retain this unconscious identity within their inheritance and the impulsive legacy this mindfulness and bodily sensitivity enact upon the present state of the world.

Symbols, afforded by the environment, become conception as the first words cuing conscious minds and sensing bodies to predict beyond the moment. The environment creates thought as provision within entities, as simply as there is nothing further to think about other than speculation or imagination, at best still embedded within the environment of self for the lack of anything else. Thought is the environment.

These symbolic words as language found in the surfaced textures of the world exist distinctively to species whether nonhuman or human: though not necessarily perceptually stated or emotively translated with any intention in equal measure or value. The discovery of physical and symbolic meaning within the environment reciprocally contours traits and niches for all species.

Thoughts remain unrestricted to a containment within the mind: this is a confusion over nonhuman and human thinking, that it becomes limited to the functions and product of the mind as calculation or solely, figuratively enclosed rather than an articulation of attached unison between each and all observances and the greater environment observed.

Thoughts as a revelation occur through needful interaction with the environment by its immersed, absorbing entities.

Thoughts and ideas are probabilistic and diverse within the make-up of reality, existing, like electrons, as potentialities until utilized. They defy separated inventions except originally within the construction of environments of reality; thereby they stay unowned and simply borrowed.

Nature recognizes genuineness. Nature expresses itself in us and nonhuman life, and with great subtlety and power. It can be temperamental, self-defacing and prone to self-infliction just as emotive compassion can occur in the human species, but to greater depth of hunger in its full craving.

The conduct of nature is the shape of our attitudes toward nature. Animals can come to believe in themselves what we come to believe them to be. Obviously, this is true of people too. Animals, entire species, and peoples can become blinded by specific human cunning and encouragement into believing they are somehow basic, nonconscious and without thought or worthy association, expendable as a fading echo contrary to boisterous human attainments found differently.

–

To walk through a forest is to pass through a deep sea of information like the molecules of water influenced by patterns of wave and current, energy and unit. Variations of this usually particle-sized information – influenced by fields sustaining its appearance, textured by the excitation of light, carried upon the air within and between the elements inviting and exciting a wide array of sensory compulsions – is only applicable to plants, and to the animals, and in coincidence.

For those species adapted to the involvement, a body of water is equally afforded information. Sensation, by whatever repeating or convergent method accumulated and individually expressed within nature, is a measurement of reality, or the dense waves of energy reflecting its features; or the temporal things it relates to as all it entangles with as a context of reality.

In this way, a forest, like any locality, succeeds among the intricately detailed, thriving sensory communications carried along by illusory fields and the lighted breeze in wondrous, coinciding coordination. A universe vastly inconceivable yet agile and delicate with its words, like a leaf playfully rotating on a current of wind and the subtle curve of gravity once the affordance for its spectacle encourages by its breakage.

Also, there can be ignorance of all or selected portions of this information. A wandering human stranger might only view remarkable displays of light, wind and air to be emotively speculated upon and become too attuned to its awareness to take notice of this other context of existence no longer of much service as survivability; or its senses might not take heed to what is there regardless, being mindfully uninformed of such textures of reality. Not all aspects of a forest absorb and reflect this information equally but by communal and residual ownership of its evolutionary processes across millennia.

In this way, the world is evoked as cause of this state of reality: a condition not existing solely within forests or waters but extending everywhere beyond the horizon of any one stranger's perspective of reality, beyond any subjective, momentary experience or conception of past and future however prominent it might seem a reason for obsessive response.

–

Nature has a strong native morality structured by the reflective coincidental outcomes of repeated experiences, and most animals live by its conscience much more devoutly than humans. When it comes to humans, animals should have to compromise this morality to accommodate or exclude us. Immediately, nature suspects us individually, and only through careful observation and proof can this suspicion be overcome as nature witnesses us as individually structured by this same morality rather than the artificial and self-calculated morality of the remainder of humanity: being only an altered semblance of this natural, evolved moral order established long before in the universe.

Also, we are inclusive with this natural morality to the extent we shape its scale and alter its evolution through our measures of impact; in this sense, we are influencing and globally reshaping these instinctive moral networks, potentially muting its cues through the erosion of the environment and the lengthy summation of our selected moral choices enacted as a resentful force upon a naïve, vulnerable world.

–

We tend to symbolize objects as implements for memory and hints of intuition; objects we give meaning to within the environment as cues and reminders meant for recognition on future encounter. Nonhuman animals do the same as reminders and cues for conduct and success, and eventually as a kind of prediction for each mind to utilize according to its prior practice, experience, and approach holding its niche allotment. Species conserve energy by using the environment's impressions and specific objects as indications for memory and response. As an example, most nonhuman animals read intentions within ongoing experiences of the surrounding environment with greater acuteness and hurriedness than humans.

The original, symbolic image is not always meaningfully apparent or retained. When we remember or think, we manifest the image, using a self-involved backdrop – our present being, coupled with the prior information of having lived our experience – as the environmental stimulus. Such connections with the environment are native and require less energy consumption to achieve; also, these connections do not require elaboration of language, itself a connecting force for the mind, as sounded and signaled among all such talented species.

The absence or nonexperience of natural or niche environmental stimulus will have consequences on how a mind will be capable of finding proper cues within the reality it finds itself enclosed and the responses required for survival based on its understanding of intention and *will* among its surroundings due to the gathered solutions of its evolutionary history.

To humanity, the universe is as we have learned to describe it. We exist on a threshold between what we describe as matter and energy, particle, and wave, or physical and spiritual, sometimes one, sometimes the other, and sometimes both, depending on our interpreted perceptions.

As an efficiency of evolution, a clan of birds such as crows might potentially share and thereby expand their understanding through a capacity to broadly learn and express that learning as a group or socially acting entity, translated through gesture, sound, feeling and any communication imperceptible to present-day humans unless closely observed or endowed with a lingering perceptional acuity or expressed genetic remnant.

Mimicking emergent solutions of nature, current human cognizance interprets as part of a grouping of conscious ideas into a general collective awareness along like lines of interest. This usually occurs with most members having limited knowledge of any overall agenda. Individuals within these groups will sometimes feel compelled, by a sense of indebtedness or honor, to take an opportunity to rally with others in support or defense of their specific collection of conscious ideas. These rallies occur equally in nature during interactive nonhuman encounters such as between avian and other avian or non-avian species within a local environment. As a commonness, birds interact well together, sharing defensive calls and stratagems beneficial to each species, again whether avian or non-avian in cases.

Certainly, in a natural setting, with diverse groupings of species sharing the same environment, they would come to recognize individuals. Time can be long and it is something which must occur together in common between species, whether through communal survival needs and measures or for play, entertainment, observation, or humor.

Whether among nonhumans or humans, such collective thinking is not responding to any truth of existence, but simply a populist grouping of ideas into collective actions within its environment. The motive for such enaction is to achieve the means of self-fulfillment of reward conceived and sought within reality by at least one individual among the grouping, even the mere rousing of a collective excitation not much different than laughter at a perpetrated or clever, occurred event.

Humans feel most strongly about nature and nonhumanity when they believe it is somehow, mystically, and mysteriously, reflecting a meaning *specifically* for them like a sign from nature, but without the recognition any such signs would require true ownership rather than gifted simply as a participant of humanity.

–

The history of humanity reads as a record of permitting unnatural irrationalities to run amuck, unchecked, and afraid of the implications of accepting a sane and realistic approach to its unsolicited predicament. These illnesses arise as divergent revolts from equilibrium and can be quite imaginative if not so power-hungry and diabolic in expression. This self-serving enactivism is usually the cumulation of choices plucked from reality during experiences and rarely self-conceived; though this is hardly an eventual excuse for a lack of culpability owed to later behaviors or absent control of these earlier willful choices regardless of source. These mindsets result from a disconnect between the mind-body and its native environment whether among individuals or dynamically within groups.

Competition of industries against the interests of humanity will be the final undoing of nonhumanity far more likely than any artificially intelligent endgame will overwhelm us.

It is conceivable there are the first and repeating attempts at solutions in evolution. Once the world returns to balance, other diverse species will emerge, but along the progressions one will advance in consciousness living atop a second's buried present, with fossils consumed, crushed, and hardened by nature over millennia.

With evidence in hand, a later consciousness will evolve smart enough not to reduce its solution to near dust. In subsequent trials, problematic aspects of the original source could regulate to a harmonious, lengthy outcome not imaginable to the anciently extinct species consumed with genetic and learned traits promoting deniability as a success.

One species could simply be the relic reminders on a course for future reoccurrence to succeed through the endowment of evidence of another's inevitable failure; or the stresses of a fading network could arouse sudden extreme evolutionary repercussions for the sake of survival. The future is full of probabilities due to occur based on prior choices or paths; yet on this path, the current one within this local space and temporality of a world, success has peaked and begun to diminish as it pertains to finding workable solutions forward in relative proportion to the loss and fatality of its once thriving diversity of species and functioning environments.

At the same time as the probabilities of success along the current path of a group of species atop this planet might seem dim, the probabilities of reoccurrence have started to increase along future lineages of regrowth and the creative inspirations for this planet with renewed attachments and symbols for species to recognize in the far distant future of reorganization.

Humanity and nonhumanity as an occurrence within reality will become symbols within the new environment for the announcement of a novel terrestrial network to potentially embrace, much as we embraced the symbols of our environment: again, once nature has grown and time has neatly filtered away the past, the millennia will advance a seemingly thoughtful consciousness to emerge, peek at its image in the void and recognize the possibilities of surrendering to the freedoms this supposedly admits, to attempt to detach and fantastically raise the reflected image above its environmental backdrop as a chosen kind.

—

It is conceivable and probable that the human species, as an entirety and on a scale with the universe, proudly knows nothing.

Human history envisions as the extended and evolving legitimization of raving enactivism, a rattle vindicating against integrity in the guise of morality beyond anything actual except by their own concoction. Words can often be much like a contagion in their influence on inactions, and equally detrimental.

For a species, any species, there are opportunities of interaction due to its bodily shape, genetic heredity, its subsequent depth of traits, and the utilization of its innate awareness as a further means of reaching beyond to find advantage within its shifting environment. This unique range of probable conduct defines the potential and capacities of a species within the current state of its evolvement: a full spectrum of common scenarios easily disclosed, with other more extreme or remotely unveiled opportunities willfully found.

Through interacting contact with its environment, the human species has carved an evolved bodily configuration as response to embryonic niches over millennia. This contact affords the species a certain robustness in terms of traits and capabilities including its use of an advancing consciousness, in conjunction with its niche shape imprinted within an environment and its ensuing disposition to act.

To pinpoint the human species within its environment is to consider the range of its probable nature as it chose to find and define through its preferred encouragement in experiencing all facets and emergences of its selected, formattable reality. Looking back at the range of human experiences as they have transpired until the present state, regarded with any brand of honest empathy, is to observe the repeated exposure of a full range of scenarios, with little source of surprise remaining to newly measure the perspectives of immeasurable nonhuman and human individuals consuming the past in waste. Humanity has proudly experienced the full, extended range of its indulgences. In this regard, imagination exhausts equal to how far the environment becomes diluted in its features and originality due to its exposure to human credibility and its physical manifestations upon the world.

The trauma imposed on all members of humanity and nonhumanity by the human species until a current state is truly, phenomenally untenable: a truth inscribed within the thinness of a sole planet's surface, one bubble among a vast celestial ocean of bubbles; and equally significant in its chosen endgame of admitting to a richly indulgent understanding in wide-ranging dissociation as a preferred, fulfilling history.

–

Just because an animal does not recognize itself in a mirror does not infer it is not self-aware. Simply because a whale never sees itself in its environment does not mean it is not aware of itself as a body and confined choice, certainly no less than the human species with its inflated notion of self-awareness. More likely, our usual self-awareness does not run deeper than a simple notion hiding an overall sense of accepted success as a dominant species.

We are full of our own assortment of cultural, historical, and environmental imprints no different but much more exaggerated than any other animal. Most animal self-awareness arises from the perspective of the outside viewing inside. It comes from the impression of other sensory and intuited viewpoints considering back. The environment and its interactions shape this self-awareness uniquely in everyone, nonhuman, and human.

Animals are usually much more than measured with dim, equivocal human insights into nonhuman awareness, feelings, or thoughts. It is our idea of our self-awareness which hides contrary realities unique with unencumbered cognizance of self and a lucidity much more associative in its self-image than one simple consciousness able to recognize itself generously in a mirror.

–

The intrusiveness of nature translated through the experiences of countless perceptions that never exist outside the bodies and minds of life except through the betrayal of subsequent actions to become observed from elsewhere. In this world, individual life does not reach beyond the current genetic extent of its species, which entangles a limitation by the environment it is adapted to as a fixed existent, unless change occurs to risk its prior success or exposes a better potential as a revised path to betterment.

To think that we superficially know the extent and depths of this vast existence; and to imagine ourselves self-aware. One individual cannot observe one aspect of existence in action without having it apparent to a potential other person in a new, closely identical wave of perspective, for only by each wave acting in conjunction, with fields and constants such as gravity and the speed of light all behaving correctly and tuned, does reality become apparent, and then only afforded by the localized limits of the perception available by each species through its niche capacities and bodily profile.

Through a coincidental truth, each wave is accurately an overlapping universe unto itself even if perceived in unison by all inhabitants capable of at least small context through perception or another intuited property of evolution given it as a trait to emotive discovery through sensation.

–

There is no imagination, no memory, but simply the recognized connections of that occurring within a locality and an entity's range of competences within that vicinity: inclusive of experiences and its evolved acuity for innovative discernment. With imagination and memory, limits are imposed due to an individual's inhibited ability to properly receive and re-connect the world; but the connections are already there beforehand ready to be re-structured, and the individual really invents nothing newer than a different interpretation of reality as it will occur in the future or has already occurred in the past. The existing individual is a constant flux of shifting past experiences and future expectations of actions based on current, changeable predicaments.

Nonhumans and humans do not attach with the world solely through an intentional, conniving interpretation. The advancing of conscious minds, an evolved method of surviving the world through better acceptance, hindsight, and prediction, is a voice trying to describe the impulses and distresses of a voiceless unconscious tendency whose true expressive attunement with the world goes agedly unspoken and unrecognized except accorded the subsequent of reality as it is firsthand interpreted by the aware sensitivities of species.

The advancing of manifold conscious minds all experiencing the same world but finding different meanings, symbols, and emotive responses than their environmental counterpart. With humans, this idea of a separation of the two is enhanced through the effects of giving the conscious interpretation dominance, as the current environment, contained within the spheres of human ascendency, trains the human mind to view the world through various well-trotted methods meant to dispel bodily feeling towards the environment as something nostalgic or pagan, but mainly as a value vastly overcome, similar to the utterance of the death of a god meant as the final separation or conscious ascendency over the nurturing intimacy of bodily or worldly closeness within the human individual: resulting in the devastating loss of emotive attachment in its relationship to a natural reality. This reaping becomes sown by an abundance of supplanted, misshapen beliefs of humanity sustained throughout history and into a present state through our varied imaginings of spirituality.

–

It is the highest innocence of humanity to pre-suppose lasting meaning upon itself. This would seem to be a comic result within reality for a species to think that its consciousness reflects a world of its unending dominance. In a self-prophesized and enacted sense, this is true, but not for the reasons projected in accordance with this human belief. Certainly, this *is* the existing universe, in all its enormity and infinite impossibility. Yet an advanced nonhuman species with deep genetic roots, conscious without perceiving the world of its design, is somehow, by human standards of mindfulness, objectified as less deeply aware or worthy of reverence for its brand of introspected interpretation of reality.

When environments change, life must adapt, alter, and transition itself. This newness within the environment can cause adaptive species to vary due to their own prior evolution which now imperils itself through changes exerted on the location. New or altered environments impose new ways of thinking and acting upon their inhabitants, since, as essential, the environment contains the potential of thought and action for perceiving, feeling bodies and minds. In this regard, in its current and past states, human beings are exceedingly careless in the milieus it engineers for new minds and bodies, including nonhuman, to relate.

–

Microbes explore the environment, moving through cell-splitting growth, beyond a physical containment in reach, to better seek survival outside the immediate locality of their bodies. Higher organisms, far outnumbered in symbiosis by microbes sharing the same extending milieu, are equally capable of this feat with enhanced range; or at least potentially this exists as a means within humans and nonhumans to attract the locality around them; and draw more from the information of reality than others can achieve through their beliefs and observations of the world as it exists to their niche instincts and bodily perceptions.

–

Throughout civilized history, we thought we were self-aware, not recurrently striving to find it. As a history, the greater volume of thoughtful words promotes a world conditioned for the human emergence, with small regard for the native ecological influences necessary for life to arise after long millennia: with humanity placed into such an exalted idea of itself as a best final solution.

In this scenario of history, nature became mined for its symbolic value while its physical and spiritual nature values as forgiven. These symbols become utilized for the establishment of corrupted uses against and as a means for the erection of symbolic power within the world, resistance becoming newly born as the multifaceted, prismed nature of these snatched symbols from a previously diverse world of successes.

It is difficult to imagine such phenomena is not prevalent, that the evolving world could be empty of such an essential ingredient. A life force capable of soaking a planet in its utterance, yet that expression is somehow not supposed to play a part in the individual lives making it up, that connections cannot form and last, genetically, spiritually, and as essential to an environment through crossing lifetimes and ages. It is difficult to conceive of an environment not being part of the life it creates, even if merely a mutual passion felt or a texture of light caught in indicating, surfacing coincidence. It is difficult to imagine lives do not connect, strongly in the rare when it comes to humanity, and unaccountable to mere chance or accident. As if nature, when a context allows it to look, does not see us as individuals and only as a species. As if the silence of its manifestation does not exist without obvious voice to hear: a silence everywhere exists as an unadorned promise only to communicate through the shared fullness of minds, bodies, and environments.

Unfortunately, with humanity, so far removed from the intent of nature in most, it becomes learned, discovered, and suffered to reclaim, otherwise the hushed potential goes on unnoticed all around, left for life to truly connect with the minimal among humanity along with other species.

Nothing lives long in isolation. A simple-celled organism cannot exist for long, isolated from its environment. Evolution stops without connection to the environment. As if a life, a species even, can exist alone, separated from the entangled world, separated from reality. Even digital, virtual, or numerical reality is still reality, still constructed of the same stuff as the rest of this locality of the universe. Nothing new originates except that which already exists in the universe apt for re-making; only the parts re-collect into something appearing new from a different, temporal perspective.

Animals can be much like us, often they do not believe what they see, and not every animal of any species becomes equally created. Animals need confidence in their environment. It is essential to survival. They experience it every moment. They sense the subtle changes like waves breaking on a shore. It is both virtual and real in experience, existing on a threshold, the merger of air, land, and water, inclusive of the forces of existence in delicate balance; where through their generational interplay, the waves break making of each wake a new multiplicity than before.

–

The mind must exist within a world where the bulk of events and information are completely unknown to it unless devised, by whatever the extended range of any individual or combined sensitivity to it.

We are products of our environment, but the environment becomes constrained by our coercion of it. By whatever lack of restriction, we choose our own production and by whatever quality.

The experiences of correlated suffering through physical and spiritual emphasis and its enhanced deprivations: these are natural, authentic reactions to a current, prior, and continuing state of confusion between a species and the world it inhabits. Just because individuals choose to ignore this fact, or express it through disordered inactions, does not make it unnatural as a genuine sensitivity to a fitting source of feeling.

Great strengths manifest through carrying these movements of emotive effect life can achieve. Without it, the world, existence, reality is just an abandoned space void of connection or entanglement. It is just a universe of empty actions. Without this feeling, reality would never be capable of communicating itself as it incredulously does with reflective insight and invention for any being, nonhuman, and human, confidently susceptible to its flow and whisper.

–

To imagine a human being is also to imagine an original version of human beings, or if viewed from a later present state such as now, prehistoric or ancient human beings, one of many close yet dissimilar species to diverge from misleading beginnings tens and hundreds of thousands of years ago, at least when considered within the context of a summarized time and space of their own making.

These early groups expanded contacts and avoidances, community and exile, superstition, and awe, furthering a native description of themselves in contact and passionately coupled with their world. From a later present state such as now, it is easy to associate naivete with these original impressions of humanity searching and unveiling uneven densities of emotion and idea within a new, providing landscape.

Art, science, and everything else become inevitably discovered through the environment. In the mimicking and later connecting of the night's stars, to long journeys of direction, homages on rock surfaces, the placement of feeling in ornaments, turning objects into possessive metaphors of awe-inspiration and specific meaning, and connecting objects into the ideas and symbols continuing to our present state of reality: conscious thought is taught by the environment giving it birth with little more maneuverability than a genetic imprint upon a subsequent offspring's mind-body development.

–

We live in an exaggerated, compounded world. Countless generations of nonhumans and humans with immeasurable consumed feeling have now quietly and vulnerably experienced its unsought existential affliction of spiritual wonder.

A certain measure of humanity is the scream of applause at being connived to believe it can think for itself, or the gaze of a stranger trying to seek a truth from another as if any truth existed solely fixed in place like an idol.

The actions of a barbaric proportion of individuals can cause the disgrace and atoning of the entirety of the rest of the species for the reason of having to absurdly submit to its tolerance due to the steadfast conditions of reality remaining from the past.

–

Humans easily build iconic structures, associations and symbols adorning themselves meant to appease and justify ruthless actions too difficult for an advancing conscious mind to admit of itself.

We come from seeming nowhere into life, growing and learning to think and feel, knowable of the fact we must all return to where we do not know we came from: a human fact of reality to generally either fear or coerce as it pertains to a simple explanation of a world's history.

Reality is symmetrical. The boundless instances of those castigated and punished by ruthlessly animated acts of the advancing conscious scream for equilibrium: or due the unconscious influence of karmic balance of forces and wills through millennia, reality within a natural world, due to a diluting loss of feeling through volume of human experience over generations, becomes the victim of this spiritually reflective conflict as it embedded its loudly self-proclaimed human counterpart as a flag of conquest speared into the center of the universe.

–

It is easiest to think of life as related or entangled consciousness when imaged from the beginning into a far distant future resembling the present state, rather than viewed along any random waypoint in alleged solitude, such as now, the present. Our evolution will naturally attempt to repeat past successes within the advanced unconscious coordination obvious to the natural world, harnessing these feats for the advancing conscious coordination of species as traits for sympathetic, harmonized unification.

Thought and the history of humanity is the evolutionary struggle for choice of direction within the species. We – as a psyche, planet, present state, as a jointly unconscious and conscious coordination of species, as substantial and spiritual, associated or disconnected – are fragmented.

The shaped environments each nonhuman and human are born into, the invention of reality, the faces of current nature and humanity stressed for resolutions, is already there upon new introduction, dampening these networks, lessening the overall acuity of the experience for the greater number. The current environment limits individuals to non-inclusion, irrationality, genuine resentments, and malevolence through an over-abundance of shared experiences beyond further capacity to clearly choose and subtly adapt.

An abundance of thought becomes persuaded by the world around the thinking mind. With the current nonhuman and human dilemma, changes imposed upon thought by the changing environment, and upon the environment by changing thought, are overwhelmingly incoherent and severed, being capable of surviving only so long as the entanglement or bond remains. Once this confidence between organisms and niches breaches a threshold, there is no immediate return to balance without long renewal through millennia. Time is a gift of evolution.

Since evolution has relied upon it as a wellspring of success, survival without the emotive connection between organisms and the succeeding environments is not possible once recognition fades. The world must begin the long process of learning to feel again through evolving a new biodiversity of species; and once felt, to think, all with the innate risk, if achieved, of an eventual reoccurrence of disobedient events happening in the future state of the world as if natural and made to transpire through the forces enacting universal and evolutionary symmetry as it is realized on this planet far distant from any other reachable as model or answer.

9. Survival of the Fittest?

Finding within a species the wholeness of its physical and mystical natures as defining fitness can be conceived as the embodiment of any individual entity's niche-embedded manifestation, its evolving, grouped connectivity of merged awareness attuned with the imbued suggestions of its bodily sensitivities, and its success of enthrallment at the overawing, commanding experiences of contended niche discipline.

Fitness is the survival of environments and the innovative diversity essential to its pronounced wholeness through an interwoven multiplicity of workable species. Fitness goes beyond genetics to collective unities of shifting interactions within conflicting, predatory, and attentive environments of excessive ploy and commotion.

Fitness is the survival and flourishing originations of wide-ranging selected, successful functions, convergent within reality as all species' aptitude for finding, devising, and understanding niches. Fitness is manifold enhancement and communicative awareness sighted through these niches and the embodying environments they jointly reap as coinciding origin.

Fitness does not fix in time and space, but is malleable, transitional, and reliably changeable to adaptations through the goings-on of its endless momentum of succeeding, roving species. Fitness is the natural cycle of unsuccessful species replaced by the adjusting evolution of others within fluctuating, moveable environments.

Fitness is the skilled maneuverability of species balancing an undulating equilibrium with compound dimensionalities of acting solutions in common as an overwhelming allowance for postured, described robustness.

Fitness is the thriving control restraining overly dominant populations from successes over the course of a world's entirety until a precariously uncontrolled present state of concentrated momentum.

–

An impulsive human species finds itself empowered to regulate and manage nature for its benefits but does not apply equal regulation in its constraints on well-resourced human populations as a wellspring of proper fitness due to a habituated denial of future outcomes. This lack of concerning attention ensued from eons of finding the limitlessness of the world and its veracious, enabling appetite for consumption of humanity's waste as enthusiastically as nature's thickening lushness hides all truths of the past.

As survival appeasement for a distressed emotive mind, fitness as a history of humankind is furthest to the centered path and closest to extremities of reality's explorations, balancing between the advancements of opposing systems of hatred to the point of overcoming the vital, artful trials of humanity's painstaking struggle for physical and spiritual resolution without acknowledging the long-ago abandonment of its further journey.

–

Survival of the fittest is not a competition of species but of niche developments and allotments, since the transformations of physicality through the evolution of species is cosmetic and trivial other than for the apportionments of traits, bodily shapes, and perspective awareness as a defined acting compilation of directed skills within a vibrating, feedbacking environment of otherwise vacated niches.

If a species loses its local rivalry, surrendering its niche due to environmental change or victimized by the overwhelming success according to the outside arrival of a competing trait against which it cannot find adaptation, the potential and actuality of the niche still exists in the same or other localities and within other future species as a closely cooperative interpretation of temporal reality convergently sprouting.

Only a barren, spiritless, exploited, and benign worldly environment with an absence the niche affordances, or consumed by unafforded niche relics, makes of evolution a quieted absence with a remote, unimproved latency of beautified occurrence.

No species dominates all others since none in advanced bodily shapes can exist for long in isolation from other species unless within a world inhospitable to all but the least of life's limits.

Without a familiar environment to slink into and develop alongside, evolution stops affording isolation from other species as niches in competition for an overall resounding fitness of an overarching environmental triumph; a solution much like the world as it transpired over the bulk of its evolving history until wrought by the growing seductions of the recently present, old world of humankind.

–

Humanity's fitness can be tricky, purposefully misdirecting, opportune, appeasing for denial, and willfully lacking in acceptance of the unacceptable. It can occur as the observance of an embraced, deluded identity of species as favored of carnal successes. Localized surrenders will become expected, but with each relinquishment of nature there enhances an intentional involvement towards future offspring trained by the winning aptitudes of the consumed, repetitive past.

Reciprocally, the fitness of all species evolves into environmental fitness, so that what was in the past becomes something new in the present as an overall spectrum of fitness for living assortments of species.

This genetically carved fitness restraining a human species to its new, obsessively focused worldly viewpoint without alternative can inescapably transform into a colony or hive-like behaviorism not originally envisioned, achieved through the straining overexertion of the genetically prescribed capacities of its bodily shape, creating unfit, adverse genetic and behavioral repercussions on a temporal scale, inclusive of near-reaching futures sullied by the favor of past winning strategies no longer applicable as sustainable, livable choice confronting a shifted spectrum of worldly, existential movement.

–

Fitness is not a singular species likened in the eyes of nature and universe, but likened compilations of adaptations, likened diversity of traits, and likened spanning, overlapping environments of shared obligations.

Environments can migrate under subtle or abrupt change to larger environmental shifts, though not with the same rapid urgency of transitioning traits among all species of plant and animal accorded stratagems of survival, and the ensuing struggle can be much like a drought of evolutionary changeover and loss of adaptability within its violently textured array of equilibrium provisioning a freshly balancing state of worldly poise intended as austere breeding grounds for all except the fittest of the rapidly-changeable fit.

Entire species become reduced, delimited in success by the accomplishment of rapid external change upon the disconnected shifting of so many environmental niches wildly panicking in unison like a silent, agonizing scream of fitted bewilderment.

The exiling surrender of a world of individuals among countless species reduced to niche scavengers searching newly fading environments for revisions to skills no longer relevant, and the increasing absence of useful skills due to an evolutionary imitation of the devolving environment. These collective species exist as if escaped from their niche cages into existential surrender, starved and choked by the new, empty reality of the scattering, de-evolutionary breakage of diversity.

Species, for the first instance and without solicitation, begin to rapidly catch up with humanity in the forceful discovery of a detached nature and the festival of conscionable nihilism as a new, unquestionable mode of truth.

–

This level of enabled dissolution exists within the spectrums of politics, economies and entire collectives of peoples exhibiting this same resolution as an overwhelming unfitness to worldly longevity and the dominating winners of one species above everything true, alive, thoughtful or emotive, as an endgame of provable waste as the scales of ecstatic consciousness, as mocking pleasures at the environment's surrender of its iconic values to human spiritual consumption for the sake of the transitory fitness of succeeding personalities, identities, and dissolving beliefs during a brief measure of celestial appearance; as an endowment of the universe ready and provisioned with the means to succeed or be overwhelmed by its far-reaching ideas.

A world can become raised to victimize its unbeknownst inhabitants. To describe oneself honestly in such a world is to describe the honors of a prey costumed inside the ready predatory traits of a donning collective personality with quantified, monetized talents of valued survivability.

Nature had to skillfully overcome all the embryonic complications of evolution for so long, balancing the convergent reoccurrences shaped by the fingers of universality, the consequences of past choices infiltrated and filtered through innumerable solutions forced into innovative continuations embedded by species. Nature awoke from a self-designed world far removed from any other in proximity or scale within the reach of light, and to far greater depth of insight and description of reality than any challenge it in turn presented to humanity and the nonhuman world with deft, imaginative, and obtainable vision.

–

Fitness can smoothly enact and enable the acceleration of an opinionated persuasion of originally unsolicited dominance upon the world as a growing force in need of revolting against as a rule of nature. Reciprocally, fitness of awareness can pronounce a counter world of empathic equilibrium, dependent upon past and present chosen paths of believable acceptance and appeasement for an advancing, weaning mainstream.

Fitness belongs to the species and individuals to predict their future accurately, acting upon it within scales of proportionality and observances of later potentials as a promoted niche dominance: a difficult fitness to support in an otherwise mute reality encouraging self-realization.

This fitness has repeated throughout history in limited spaces gradually coming together and joining into a single fitness ripe for directed solutions of its own contriving insights. This diversity of species became fit to formulate a world attuned to its iconic marks of self-interpretation in sequestered absence from the quieted universe, absent from minded contacts with forces of life minimalized by one chosen pride marked as primary and fiefdom over a worldly entirety.

If success comes from the conquerable, there would be no diversity of the natural world. Nature would have already found this solution. The belief of an alternative success begins from the original, temporal mistake of an advancing consciousness, while reciprocally a random species, to mistakenly claim originating solution to a world's lasting future as the new, chosen arrival manifested.

–

There are individuals to see the feeling of existence and consider the futility of striving for dominance within a reality, of trying, caring, searching, and enduring for the sake of a supposed, illusive meaning. They feel the barbarous intent of an unobserving, empty reality founded on nothing more than fabricated perspective of a world hardly penetrated in its unequalled depths.

Finally, after enduring eons of evolved agony, these individuals arrive to find a world asking and teaching the question, with sympathetic inflection, as to why a biosphere would strive the agonies of evolution for eons to produce a singular species for its spiritual meaning in the otherwise emptiness of a cold universe, to become this means to an end as an instinctive, amusing wonderment at survival's final product.

Surviving because of a landscaped reality, established as a surreal reality of identity as both master and slave together like particles and wave permeating humanity's reassurances, without the primal peace among species afforded by nature's prior forceful boundaries, nature does have its tolerances, rewards and detrimental pitfalls accorded the designing ambitions of any singular species.

Survival of a species to no end other than the idea of its own dominant physical and spiritual natures as servicing justness and self-claimed enlightenment for the sake of no better aiming invention of imagination, scrubbed from the prior environments of succeeding betterment, humankind has supplanted the integrity of natural subsistence in favor of the instinctiveness of economic survival as a divine sacrifice attained through excruciated flailing of the originating, spirited nonhuman and human realms.

–

Human aptness opened new avenues of exploration for complex environmental fitness, for new potentialities in combination with a dexterous bodily shapeliness surfacing from the multifaceted world of unconscious, attaching encouragement; perpetrated instead into a wildly celebrated, rebellious, primal scream at an attempted echo as response, a familiarly self-willed, unanswered resounding as an otherwise mute proof against celestial certainty.

For a singular species existing on a planet without capability of reaching another similar, alone in a universe without direct contact with anything further within an infinity other than streams of venerable light, and allowing its world to deteriorate, for the manifold species of its planet to swiftly exodus from existence into dust, for future generations to miss the opportunity of existence on exhibit atop a planet in an existing, awing universe, would seem a reasonably questionable sort of fitness newly explored as an ambitious origination of innovative reality.

10. Spectrum of Depthless Enthusiasm

Envisioning a Normal World

Attempting to imagine a normally intended world provokes the obvious question: what is normalcy or the normal? Normalcy could interpret as a genetic emergence of well-adjusted, abundant nature and confident, niche-bursting environments, or as social clarity within a self-contending, self-aware species privileged by an endless, unknown universe it loudly proclaims found and encouraging to the reverence of its kind.

Normal is genetic distribution as a memory of cumulative solutions of the past: a method of extension connected as a lifeblood to the docile properties of reality through regulated reworking of its fixed forces to advantage. It is the same as species learning functions in nature through the influences of perceptual beliefs on future emotive, bodily explorations. Within evolution, it is the search for betterment through solutions, the normalcy of survival within a created atmosphere of heredities and embedded, temporally enacted enrichments.

Normal is the perpetual, transitional reuse of an animated, evolved reality, and the continual surfacing of the present moment as a developing culmination of the past. It is light, wondrously absorbing and reflecting a vibrant reality of coalescing forces for life to surface and affix with advancing evolutionary explanations. It is gravity gently coaxing a speck of dust to the ground while held buoyant by the friction of soft, stilled air and the exchange of contact. It is dust, the smallest exhibit of humankind's unenhanced observation, far-reaching in its inner forceful, complex and particle reality, intimate and vital to the impassioned, stimulated observer while occurring mostly beyond any spans of accessible attention.

Normal is a universe incredibly grown into the synchronized forces of existence within which life and consciousness might appear to ponder its fitted place within it. It is a diversity of mutually shared experiences and innumerable awareness bound by hereditary solutions over a spanning landscape.

Normalcy is common ground mutually and temporally within all species since a distant beginning when a recently initiated life attached itself and began traveling the opportunities the mentoring forces of reality newly allowed it to divulge, with its puzzling, effectually textured surfaces to affix rightfully in the middle of an infinite nowhere within a newly disclosed creation of regulated imagination.

Normal is not actually finding this universe except as an observer of something which might never become measured in its totality: an observer barely comprehending the local setting of its personal beginning, identity and making beyond the utilized bodily and environmental tools of survival it has received by an affording world and contented cosmos.

–

An observer of the world might become normalized and designed by its evolutionary choices and the solutions it is biologically bound incapable of viewing or truly imagining the hidden workings of reality except incredulously as coincidental accident or wrought as a reshaped iconic discovery plucked as a colored petal from an associated environment. Normal is the behavioral trust in changeless self-interpretation as the conviction of survival; and in the depriving promise of further insight and sensitivity still to occur in a furthering, impending future.

Considering normal in the natural world obviously includes nonhumanity in all its eloquence as an innovative mountain of multidimensional bodies with emotive motivations, heightened sensitive and interested awareness. Humanity in its current state is a recent occurrence within and outside of nature and should not have been capable of enduring the complexities of its evolutionary bloom without a hitherto established level of normalcy in genetics for stowing behavioral memories for countless improving species. This abundance of animated solutions became designed into the living environments they shaped through soliciting answers atop an introverted planet, within a boundless universe often misconstrued as the divinely inherited ownership of humanity.

Supposedly and permittable by an unrepresented, unwary, and unguarded nature as a choice of explanation within reality to implement, this ownership allegedly permits its full consumption for the attempting benefit and growth of dominant successes for one winning species designed capable of its truer refinement accorded its capacity to overcome the natural world for an advancing winning posture.

Humanity would not exist in its current state without the established diversity of memorized solutions shared as a representation through the genera of life enclosing a planet in its rapt interpretations; and occurring within a universe as a normal, bounded state of overwhelming, willful attainment not intended to be ousted by a singularly provoked, overcoming species as a genuine truth of encumbered survival.

In the world of genetics, convergent genres of skilled traits within species, niches, environments, perceptions and enaction, all with complex emergent realities, there is range, deviation, unifying multiplicity, the merging and distribution of enabled solutions bound to the fixtures of the existence they evolved alongside, as well as complexities of yet unfound probabilities of answers, pitfalls and traps awaiting inside reality to be unveiled and sprung into the unsuspecting world as the compulsive drive swelling from the struggle for better universal elucidations.

Out of this same world appears playfulness, curiosity for its own sake, innovation, and creativity for all species within divergent perceptual abilities and mental feats of invigorated, opined mysticisms.

Normalcy is balance, its fluidity observable in all facets of reality, in human-made science and religion, social politics, economics and overwhelmingly aligned alongside the natural world with its values and enforced fetters, and in the ongoing saga of human evolutionary confrontation with this poise of multilayered life through its sequential epochs of conflicted history.

The ebb and flow of eventful, coinciding histories reciprocally developing as betterment or deterioration through risky divergences from this illusive equilibrium or middle ground is the forceful struggle for a constancy of reality. This symmetry must exist in measure regardless of the actual perceived values of universal forces even if uneventful emptiness were the sole ruling force of the universe on a zero-sum scale. Without the vibrating deviations of a current existence permitting the affordances for advancing realms of extremism within the wide-ranging cosmos, there would be no possibility for life to surface as an otherworldly experience floating on a solar, frictional projection through an inert, benign emptiness, an only technique affixing physical manifestations, and as an exclusion from an otherwise alternative stillness.

The underlaying delicacy of a symmetrical world made for diverse changes to merge into evolutionary paths, for the movements of life as an initial twitch to fatefully find and manifest as fruition within a flexible, shared, delimited and successful lifeforce tendrilled as countless temporal species bursting with bound genetic designs for the continuance of its allotted behavioral realization, the constancy of its flowing waters and energizing tides, its colorful, weathered and appeasing appearance, and its mountains of buried, hardened antiquity: a hard-fought multiplicity of answers permitting to occur the wilting current and previously wide spectrum of species as a collectively focused, cognizant state of reality on a planet.

Normal is the observation of equilibria at work in the natural world, as the innovations of survival for all species to disclose as awareness, valuations, and skills of actions through the extremes of bound artistry reaping its predatory descriptions upon the suspecting, deeply enthusiastic and unbiased world.

—

The spectacle of light from the continual rolling occurrence of the passive rising and setting sun on a rotating globe by a conscious species rooted to its place: endlessly, a circling maelstrom of locality is exposed to an inimitable sun's appearance and disappearance while capable of being deliberated or celebrated, temporally joined with every other spot on its surface based on the multifaceted, bloomed forces shaping its minor twirling occurrence within a nondisclosed speck of the universe. This spectacle of light and ledge of reality belongs to all species as a beginning and a budding source for uncovering symmetry to traverse the paths of life.

The original normalcy of reality and the natural world of species and forces can be irresistible for a newly conscious awareness to find itself attentive to as objects conquerable to its reachable courage; while held fixed to its originating locations as furthering, unravelling stories to widely expound.

This abruptness of discovering itself as a distinction within a universe of which it knows almost nothing except its primal, genetical skills of navigation making it environmentally aware, yet mindfully unaware of the eons of time still to come assessing its qualities and meanings just as it has for expired eons of the past.

Normal is the alleged real world, the world of nature shaped by a celestial realization beyond the scope of any one species, cognizance, solution or belief as proof; rather than as the singled, attentive focus of an omnipotence, as if the infinite cosmos were simply the meager summed likeness of a shining, gilded light casting downward as a spotlight on one planetary stage.

Species would recklessly and futilely attempt to reach a captured environmental world of niche successes such as on this planet without this malleable spectrum of depthless equilibrium binding species together as a normalcy of biological, physical and chemical reality: explicitly culminating in the relatively recent past, arguably peaking at the beginning of its regression from fitful successes with the advent of humanity a mere tens or hundreds of thousands of years ago as a newfound, unleashed, fleeting ambition upon the lush natural world, consented by nature's obvious lack of impressive resistance.

Envisioning a Surreal World

Observing the succeeding system of a normal world practiced within the ecological circle proven by nature provokes a new enquiry: what, if anything, as a state within a collectively aware conviction, is surreality? As a concept, it is perceptual abstraction of reality as it exists in a current and past state, not much separated from hallucinations, or the upsurge of the unnatural and the embracing of manifold aftermaths within the complexity of reality's strangely unifying alternatives. It is an evolutionary shift from that which proved itself over spanning epochs of millennia and contested through comparatively short generations by a defiant, asserting, and ambitious progeny.

In this plausible surreality, interconnecting minds not only envision, predict and construct niche fulfillments as the opportunistic boon of genetic successes, but prove a world beyond an originating niche comprehension or intention of design. The surreal is the world of emotive environmental connectedness turned into fantastic, magnificent ideas contorting natural awareness, of proud, collected survival transformed into an imitating individuality of worded honors, as the obligation of illusive denials and odd beliefs as enlightenment, and as the malleability of emotive attunement into a plausible, complexing origin within humankind as the solely represented, uncontested challenger within the universe. The necessitated condition of surrendering instinctive and heroic belief for the appeasing explanation of mortality for impetus of action is the bewailing surreality of a newly raised, political world.

Surrealism is the unreasonable idea of meaning thrust upon humanity from nowhere except the exit from the jungle of life and embryonic reality, with the caressed, concocted idea of each subsequent generation holding to the belief of the exodus as completed, its status properly translated, and the distortion of this apparent meaning with and from its worldly origin as a privileged, just truth to be found and lived by for the furthering of human pride and its predictable, impending radiance upon either this world or the otherworld.

The old skills of nature reshaped, retooled, and relentlessly readapted with unacknowledged primary aftershocks to the minds and bodies of human and nonhuman environments. This new skillset includes a vast range of unpausing past and occurred extinctions passing over as normal and without pangs of regret or the much-sought integrity of real, valued importance once extinct from the sketched memories of human observations.

Humanity enters a new, surreal world of its own founded atop the old reality just as life overcame the prior barren, inhospitable landscape of the early planet with the same imbalances of vigor.

In observing humanity's surreality, it interprets what is normal in a world modeled by averages and majorities as guiding principles, by unfounded opinions and quirky moralities, and where truths are speculative prisms for the success and ruse of ambidextrous desires. Normality as a fully existing average or point could only occur in an unchanging, unevolved environment, one either never begun or fully concluded without potential for a further inserting solution to disturb the system into an abstract, abnormal state of discontinued deregulation.

This sense of normalcy is the spontaneous world of human surreality, like the first peripheral absorption of environment by life inevitably consuming a planet in its diversity and shared diffusion spread by the branches of evolution, with long-lasting generational space to cling to for support as an archive of solutions.

This method of forgiving an old world could alternatively be applied to growths of species and transitional solutions as native predictions of reality and the regulated variations of generational qualities, much like the essential range and advanced scope of diverse and convergent trait development in conjunction with niche environments, without one niche or species being the epitome of normal or origin, but of mutual forces embedded within the framework of life as a fountainhead of mutually attained actuality.

The is no energy or might in isolation, starved by a barren, dissipated reality erased of solutions and predations through the absence of environmental stimuli or the degradation of a supportive landscape for life other than humanity to attach.

The world of surreality is a world in which its wide absurdity is noisily pronounced but mostly unobserved and unfelt, and where the natural repercussions of such choices are normalized into appropriate behaviors through prefabricated, integrated systems of belief as ready choice for newly arriving participants to learn and succumb to admit without opportunity or alternative to its surrendered, closed state.

–

It is difficult to define intention, an essential property to be skilled in to correctly discern for better survival in the natural world, and how it became sourced as cause for action within a surreal world. Inclusive of the revilement readily permittable to nature as a lower rank, the ongoing successful dissociation must become learned within the same species by biased division mimicking nature's systems of change-encouraging, contending predations.

Individuality within the world increases and shifts from clans as knowledge and experience grows in associating specifics of environments to predicating memories and the roots of active opinions. When the world integrates newly discovered inspirations and individual-based inclinations into the traditional intentions of clan-based drives and the intermixing of a diversity of group-founded beliefs of the same environment as an inevitable outcome of growth, conscious knowledge of truths become broadly conflicted by the originating environment of a solitary planet within a vast universe of immediate, original ignorance, as knowledge first begins to develop and gain symbolic retention to solid bits of cultivated information.

The improvement of retained conscious information for an uninhibited species, or the long retreat from an original illiteracy as predetermined by the fact of having an origin from which knowledge must begin, is not equally spanned or far removed across all individuals of all temporal states ranging an immensely affecting and overlooked current, prolonged moment of the surreal world.

The advent of original ignorance and its impending, evolving errors upon a calculating consciousness cannot be understated as an influence upon its proceeding comprehension of knowledge, no differently than its ageless pursuits of instinctual values during an original, niche-fathoming environmental world allegedly disclosed merely for consumption, as the inevitable walking stones provisioning the glutenous path to humanity's enormity.

Lacking through the history of this accumulation of environmental information over sequential epochs is the differing degrees of misunderstanding and disinformation, the entire spectrum of evidence as it exists to misleadingly confuse reciprocal to how it enlightens. As difficult as it is for conscious minds to find association with the obvious, realistic side of information, it is exceedingly more difficult finding the sources of disinformation's unsuspecting deficiencies outside the awareness of current visions of understanding. This arises due to a practical ineptitude in knowing what is further absent from satisfied conceptions, targeted opinions, and the scheming resourcefulness of predictability in its association with information previously wrought from reality into new contexts of eventful coordination.

This dichotomy of confidence is like being aware of a universal world through the inbred lens of surreal networks of acceptability alone, and the lost fondness for the natural world long considered offensive as fitting source of true empathy, or as further deliberation of spiritual meaning beyond incidental fodder for human physiological and ascending sustenance. The natural world is nonconsensual in its symbolic uses as essential, usurped betterment of solutions for humanity's inevitable consummating ascension as the winning, inimitable personification of everything a planet has to offer it.

In a current state of humanity, it can be hard to appreciate the power of original ignorance, even affronted by its thriving durability as it presently exists, upon the unawareness and inexperience of a budding, virgin consciousness as it first explored and began to find the wellspring of environmental knowledge, as the derivation of correlating symbols, as a beginning, empty of anything which exposes itself as closely reckoning it's meandering paths from instinctual experiences to the knotted strings of surreal learning.

With a suddenness transpiring over millennia, the human being is abruptly stated and duly recorded as highly conscious, self-aware and privileged by its starting witlessness, accepting the repercussions of all the emergent scenarios it will inevitably confusedly exude as a symbolized, surreal history of humanity indelibly written into the fabric of reality through its normalized acceptances among winning patterns of flouting: dissociating itself cleanly from the prior integrity of the world.

A private, imaginable world is suddenly possible where it never existed much before. There is a valid absence of culpability before fresh truths supplanting old ones, with methods of measuring personal space and alleged freedoms of private awareness in the company of strangers all capable and deriving the same mixed, befitting, and regulated ends. This privacy does not even have to exist in isolation but can share in associating experience by others just as eager to articulate and overcome its dizzying opportunities, to reveal its physical and mystically held promises, and to divulge its latent, appeasing realities to one another in celebratory ritual.

Life on a planet in this instance becomes a game for the world of humanity, a stratagem of *us against them* and *me against you*, even if some strategies seek transient cooperation or submission for its better survival in the precarious and consequential moments of long eons of history behaving much the same in innumerable, differing localized, overlapping and convergent contexts of history.

In cases, relative temporary isolation can promote values incapable of a jointly shared world due to the absence of palpable measures of risk as an outcome of its restricted approachability protecting against conflicting opportunities for elimination, creating a perilous future due to the unwitting fact there is no lasting isolation cast as eternal success within the structured designs of its world.

The same could discern for a species isolated within an unapproachable universe and its conceivable capacity to share a singular world bordered by cultures, shared as the mutual venerability of all species alongside nature as a hailed unifying journey, with each burdened by redundant perils to outwit.

The branching future outcomes of any challenge upon the dominance of an otherwise conceivably impossible fact of appearing clung atop a rotating, curved surface might appear endlessly unfathomable, other than for humankind making of it a perfect refuge for the newly discovered identity of an individuated self to generationally advance in play into an unsolicited, groping and sole choice of evolution for new participants to unquestionably follow without alternative.

This contest within reality for supremacy translates into an exclusion from the preceding, cosmic world as soul-filled substantiality to a purported, inner, human-specified world. The new unseen world discovers to offer an efficient motivation for encouragement against the nonresponsive celestial backdrop exposed through nature's inability to properly defend itself against its losing roll of the dice. By its proper actions, humanity becomes absolved of its effects by the rightful cause of its surreal rewards of redemption.

—

Among the synchronized contributors growing a viable species over millennia, a real world can appear vastly different simply due to standpoint, the delicate traits and cultures of followers, and the idiosyncrasies of the manifold environments as influence upon its willing capacity to become identified entities. Within the scattered realms throughout history, humanized niches became found, wrought, and defended no different than the rest of nature and its niche-satisfying, conjoined environmental medleys of self-wreaking genera, as species have recurred its defense for a relative eternity of years.

Something observable in one mind is unobservable in another simply due to a differing sensitivity of indications willfully associating the makings of reality, due to subtleties of variation among members shared genetic construct, pragmatic natural encounters, localized reoccurring traits of species and eccentric mysticism.

These variations are natural and considered normal for such a species as its means of surviving diverse external risks towards advancing solutions while still supporting the signatory identity of the species as a collective. These variations involve differing patterns of perception, making entire collections of individuals of a species capable or incapable of manifold posturing of mind and control of body within reality based on genetic prevalence within the whole by its inheritable proportionality and cultured emergence into reality.

A suddenness of appearance should be a consideration of human resolve: as if we do not realize as a species the abruptness of our newly conscious arrival within reality and its contended meanings.

There are deliberately misguiding perceptive beliefs of this emergence as a surfacing of the human-made universe: needed to exist as a workable, long-standing solution for surreality to best persevere in the tendencies of humanity's future meanings.

Aspergic Tendencies in The Surreal World

Self-confident, personal incautions within a private mindset contrary to the usual interactive, selective tendencies of a preset social forum can be dangerous to possess in the surreal world of humanity even if they can be beneficial in the real world of nature, creating a fractured subjective experience in its doubted reality for the observer alerted to such a riven state of introspection.

Focus mistakenly perceived as attentive absence favoring a distraction while preferring the secrets of otherwise overlooked truths can create incautious risk where social devotion inundates from a crowded world; and where malleable truths can become altered or conveniently fitted without respect for reciprocating penalties stretching beyond the servicing of the moment.

This ripened shortness of integrity before the world can be unquestionably strange from an Aspergic perspective. There is the cautionary acceptance these peculiarities of mystical answers and perverted justifications of actions can be learned from the outside world and thereby must be privately retreated from as a passioned commitment in life, personified as a sedated revolt against conformity to what should be considered perceptively and sensitively abnormal even if benefitted by a hailing multitude of a species as a well-donned victory.

There is a self-serving Aspergic resistance to taking on these crafted personas of being, rather than behave naturally, though this naturalness is unknown or mostly lost as a plausible adoption. There arise clearly understood reasons for these personas in plain sight, even if not fully believable and highly exaggerated beyond an Aspergic ability to other than internally revolt against as motivating paths of experiences meant to cause future resentments: while existing empty of any predictably lasting ambition as a target for humankind to generously muster.

–

If exclusively accepted as believable through the stimuli of mounting environmental experiences, as reparation for what is lost or denied with equal capacity of penetrating vision of the surreal world, there is a high acuity for Aspergic association within the natural world, for a shared emotive curiosity, for beliefs in patterns of predictable happenings, and from those patterns the disclosure of the hidden world of reality.

It can be revealed to be in plain sight all along and vastly ignored as it originally thrived beyond the histories and affiliated unifications of an interacting humankind as precisely occurred and unconfused by its successes and realizations, while consumed with lasting insightful awareness not much beyond fantastic, momentary superstitions of organized, cherished mysticisms and the acting ideals of imaginable heroisms meant to reflect grandness as an iconic confidence.

This historical, lacking authenticity creates a world where an attached belief is a preference for *something which is nothing* as an idol of reality and motive for actions rather than merely *nothing at all* or having to face admittance of the potential void of spiritual existence without assistance from raving feral imaginations found through original strangeness. The world readily offers symbolic meanings and attachments to any wayward, insightful attention ready to corruptly foster into answers for populist use as mindful absorption into a wily preference for unfounded beliefs.

This is a surreality in which the spirit of nature's life is wisely silent, consumable when placed in context alongside the broad successes of humankind's canny spiritual revolution. This enduring spirituality exists within chronicled participants of times and locations, unavoidably shorted in will beyond a singular insight to observe and decide its world, with an unpredictable future in need of nurturing a cultivated appeasement as its dichotomous duality nakedly exposed before a deceptively watchful existence.

The recurrences of nature, its cyclical, reenergizing transitions of departure and reappearance arouse an enabling sense of confidence for the tendencies to Aspergic comforts promised to searching minds, and found illustrated through life's sensory, stormy attractions. There is comfort in existing as witnessing observer to the moveable truths perpetrated at a scale of rolling permanence organizing the fruition of possibilities, as likelihoods of further occurrences to follow, and as intrinsically raised, conjunctly designed by the skilled, fingered dexterity of universal forces.

The natural world is more recognizable and identifiable in its unaffected replies and usual predictability. This genuineness does have residual seepage observable in the surreal world of humanity as well, in instances where nature or a natural reality is observed in experience with an impassioned, ephemeral detection and the entirety of the world is shortly exposed as still living, composed of material, substance, texture and essence beyond the normalized human exposure of chosen limits of sensitivities, of closing endpoints of understanding, of illusory trickeries of senses, bewildering connotations of truth, and prospering beliefs in a newer, grander realism of self and the world without vital appreciation of a contextual reckoning of proportionality embedded within the curving, yielding designs of universal scales.

–

Expectations of authentic outcomes when dealing with the attached opinions of commingling groups of individuals within a surreal, concocted world is an expectation of overcoming it and purposely inhibiting the vulnerable mind composed of Aspergic believability. Within crowds, there is seeking dispassion as the best means of protecting an essential impassioned identity as nostalgia for the motives compelling undeterred individuality.

The crowd exposes denied, misleading talents within the complex routines of societal enactivism. The Aspergic personality can become absorbed in distraction, lost in self-engagement within a crowd of reaching ideas and the promoted insertions of wildly divergent, motivated, influential exchanges of experience and ideas of truth. Truth loses its aura of authenticity as it becomes a benefit unique to individuals while dynamically grouped to localized cultured opinions and forceful, exaggerated credence.

–

Aspergic tendencies can establish a lifelong process of rebellious, reciprocated self-learning and self-teaching, whether the lessons taught are from oneself or insightful others during watchful experiences seeking new, keen-sighted inspirations marked by patterned, humorously strange and unexpectedly connected presences. It makes an individual believe in a perceived world which exists better in the enaction of others, while the real world of behaving, sensing, and seeing a different texture to reality becomes an alleged fantasy.

These fantastic insights dimly meld into a world of alleviating entertainment and the organized consumption of instinctive essence into a differently coiled identity in its association with absolving the excruciating stamina of time. This detachment becomes learned through experiences of long lingering deliberation usually to utter exclusion as a persuasive factor on selection within the realm of human surreality to which it is not precisely related primarily due to its fitness deemed as inappropriateness, untranslatable, or not of a proper kind.

Confronted by resistance and a sense of defeating self-doubt at an overwhelming, isolating mainstream, distrust inevitably will affect decisions, in oneself and in relationships with others due to personal inhibitions mesmerizing the normal spheres of popular plausibility.

Doubt emerges from the sense of being so minimal in relevance that one solitary viewpoint must be benign in context with such celebrated popularity, especially without the resolve to observe with appreciation divergent experiences of mind and body in conjunction with the stimulus of accepting the preestablished, affecting worldly environment. This current environment collected and endured over eons, purposely structured into the proffered reality as the one welcomed into by an unsolicited rite of individual birth among its recurring lines of unproven, alternative successes, organized to warmly embrace its first and subsequent moments.

Overcoming the world enough to embrace this resistance can produce great benefits, reciprocally for the individual and humanity it might immeasurably inspire. Subjugation to this resistance can create a disheveled, conflicted, and soothing solution for the Aspergic individual, and a comfort for others to dismiss as a better fitting representation of how a shockless, mystifying normalcy is supposed to be carefully imitated.

Normalcy comes to be a loss of control expectantly found yet not fully acted upon to avoid hardening a lasting replica of identity. For the Aspergic attention, resistance itself becomes a means of control, not as a stepping above others, or as judgement necessarily directed at other standpoints, but as a simple method of finding support in oneself without loss of identified difference. This sense of identity becomes compromised by repeating recurrences of reality exhibited through the seepage of emotions conjoined with perceptively sensual sensitivities exuded during human interactions impossible for an Aspergic awareness to ignore without simply not looking or not giving direct attention to it as a method of distraction, and with a confident persistence like pride.

Aspergic attitudes perceive the surreal world as a fusion of past and present, a happening of temporal disassociations difficult at times to detect with sustained willpower to watch its reoccurred histories. They must recognize to contend with the fact they contain at least a measure of the assumed absence within *surrealistic* observations when it comes to accepting believable foresight, like an equalizing constant of mathematics necessitated by the origin of personality and its requested validations of appeasement at an expectable social product.

In the observances of surrealistic adoptions, this subjugation eventually comes to be a directed expectation furthering the overwhelming opinion of normalcy. Normalcy becomes repelled by natural societal resistances. These repellant reactions are often not much separate from fears of unawareness, of what is unknown through surrealistic experiences and their envisioned ownership of the world's bigger, physical picture without the loaded subtlety of exhaustive undertones.

To an Aspergic personality, surrealistic normalcy can be an enthrallment contrary to a naturally minded quest for equilibrium as a relationship with all reality. There can be guilt over one's own social inadequacy. Inane separation can come from not wanting to impose such great exertion requirements on most others for the sake of a singular attending identity.

There are many real capacities of furtherance the world does not observe except through the focusing lens of Aspergic and other non-surrealistic perceptions, including the deprivations of nonhuman and human sensitivities as premeditation; at times, only later in life do these self-requesting personalities become alert to vital adaptations to surrealistic facets within the everyday world of normalized surreality through a discreet, nondisclosed relationship of endurable unification with it as an evolved, private adaptation.

Aspergic personalities persuasively learn they must adapt to whatever others resemble as personality, intention of opinion and willingness of action for a vaster collective popularity but struggle with the implications of approaching evidence observed to be naturally worse on the integrity of the world than to adapt not as a mimicking of others as sameness but as an imitating likeness attempting to preserve a minimalized involvement, and thereby a less convincing temptation.

This is not easy as a choice for an Aspergic attention to accept as the condition of social identity within the realm of humanity due to its overall lack of prior and current credibility given by its generational shortcomings. There is palpable evidence of its believable erosion of attention realized within the observable world of practiced humanity not realized within the observable universe or the innovations of nature as elucidation.

Surrealistic minds will often assume a required, intimidating adjustment due to a strange personality being isolated and labelled, as an imposing, expected requirement from that other, observably unresponsive, intimidating individual to clearly sense. It is self-appeasement, and an unwillingness to learn and adapt oneself, which strongly motivates the surrealistic mind full of its own vast assortments of motivated behaviorisms more damaging in scale and likelihood of experience within the world, all considered collectively acceptable as recognized talents of what properly defines normalcy of active attentions, however limited or overcompensated in scope of imagination, imitating talents and the command of personality in calibrating context while riding the unfettering waves of surreality.

–

Often, though scarcely recognized without prewarning, an Aspergic person can be a circumstance of enaction for surrealistic individuals not otherwise found in the experiences of everyday affiliation among a mainstream of duplicating individualities, encountered as a selective opportunity for openness or suspicion as an austere persona, to uncover a seeming eccentricity as either interesting or as a personal affront to the well-worn uses of traditional acceptability.

These sporadic opportunities will not be equal in outcomes, due to the mutuality of the tangled, pre-established dissimilarity in surrealistic perceptions instead commonly and misguidedly interpreting the Aspergic personality to be a haughty exaggeration of normalcy or simply too challenging a discrepancy in perceptual readability beyond the urgent spectrum of surrealistic normalcy to believe, or unvalued of further chasing for its unique resolve. This disinterest can reinforce an emotive separation experienced by the Aspergic mindset, further enabling the recessed appreciation of social dissociation and the internal depths of a marked, isolated identity.

Aspergic tendencies observe the paths of balance within the environment of existence as an irresistible desirability, enthralled by the detailed implications of an unconflicted, genderless nature and an existence demonstrating an introverted caution, a simple complexity and an unwavering, embracing confidence in wide-ranging transformations being inherent as a movement to the structured frolic of reality.

During social interactions with and observable between people, it is the recognition of one or an assortment of idiocentric traits observed, preserved, and re-sought as pleasing repetition or rediscovery, of related preceding moments occurring as a future memory finding entangled context during the happenstances of the rolling present experience.

Aspergic belief can be an acute relationship with reexperiences during the occurrences of current experiences.

Aspergic identity pursues the gifted sensations of a bodily mind, aiming to substitute for the same or better perceptions as any other unbiased person, mutually assigned and existing as a mere potential connection between the happenings and the dynamics of the unavoidable human encounter with its own species during a lifetime within the pre-created world.

This makes surreality for the Aspergic inquisitiveness an intended, alternative world in its usual absence of stimulus other than, at best, unattainable idyllic or idolized inklings of normalized, mystical unreality as the inspiring fodder for reaching remotely comprehendible human spiritual realms.

–

There is overabundant exertion attached with retaining the past as an inner working mechanism of connotations, a playing field of self-analysis, rather than a dismissively selective approach to the retention of personal memory in its association with the world, as external identification within societal milieus and its cues for an identifying personality to be proudly done.

The demands of memory retained as strict replication can be confounding to an Aspergic perception, especially without the direct appreciation reciprocally gained and lost through a normalizing genetic diagnosis or causal behavior patterns as familiar clarification which would have been prevalent until the recent past: whereby Aspergic individuals lived without knowledge of a difference in their perceptive insights, assuming the distinction an error of processing against a vast array of contrary personalities. In another context, it saved these individuals the victimizing shame of becoming labelled as exceedingly broken rather than simply overlooked, if not wrecked on exhibition in a personalized asylum. Thereby reality creates, through constriction of alternative identity customized as forceful conformity to escape, convincingly self-identified persons in contrary belief of a personal normalcy.

With the requirement of memory retention as a method of associating past, present and probable futures, there is Aspergic difficulty in letting go of the billowing meld of personally salient past events as pointed markers for structured use later. Aspergic personalities can come to cravenly seek newer experiences for improved associations of reality as it is properly actualized.

This occurs without much recourse however alienly conflicted an individual Aspergic persona might be in comparison within their own emotively thought experiences as they try to grasp the eccentric world in which they find themselves exposed to while juxtaposed with a composition of motivating, externally composed worldly intentions beyond much Aspergic interest or true reassurance as ownership.

Non-Aspergic people, by their own sketches of normalcy, are different through surrealistic tendencies while existing in commonality. The Aspergic person exists within separate parameters of perceptions and feelings of thought in response, often without realizing until later the parameters are different, while finding common imitation within the surfacing landscapes of existence, creating an internal person in conflict with its own human-made environments inclusive of massed, perpetrating identities: unless exiled to an intimately sought refuge of private self-making.

This straying departure allows for new insights and abstract approaches due to a demanded ability to view the world in other plain dimensions of evidence outside the bounds of surrealistic human normalcy.

The retention of strict memories also applies to human histories, where the past is not easily or even possibly overlooked as a source for higher truths, repeating patterns of behaviors as explanations for the past and current state of humanity within the world, with the passion of a nourishing elucidation others do not necessarily know the compulsion to seek and the confidence to admit.

In social interactions, an Aspergic-leaning personality can become possessed of an imbalance in responses due to a retained relationship with happenings through heightened memory and the reexperiencing, newly associating sensitivity to the impulses of memory during the moments of occurring encounters: more as a convincing fondness for focused dissociation than a disordered conduct. This experience can be heightening when experienced in conducive isolation, surrounded by preferred events, or likable settings and gatherings.

Aspergic tendencies create an observer, a seeker of occasions for latent delving without having to be a direct participant. The Aspergic person becomes the cause for the effect on normalcy and the mainstream it imposes upon as an odd, discomforting suspect. If the two interject calmly there is a measure of sensitive phases of normalcy within the array of personalities making up this state of unfinished experiences. An Aspergic identity will correlate experienced attributes of the temporal and spatial world not looked for or suspected by the normalized world of fixed, believable harmonies.

Reality is a lure, its textured, thinly coated truths readily plain to accept in an Aspergic personality must be arduously striven to acceptance in the normalized mind excessively devoid of Aspergic tendencies as a clear, gifted abnormality.

–

Aspergic belief is unfettered trust in its recognition of a post-truth and post-ethical world traversing the depths of events and experiences back to a near origin, while questioning the validity of the prefix "post-" throughout human history as a context. It appears more as a constancy or more proper to the prefix "pseudo-" or "anti-" in its relationship with the evolution of humanity's collectively aware acceptance of surrealistic identity within its reworked surreality. It is a lack and confusion of human history which transcribes the prefix "post-" to something which appears new but is truly ancient and has always been since a far distant infancy of conscious belief in its own tarnished, transcribed supremacy.

An Aspergic person can come to empathize with nature as an association of marked context and reverence in status associated to a contrasted dominance of otherwise surrealistic overbalance among humankind.

–

Though enjoying the strategy and thrill of games, Aspergic tendencies might not enjoy taking part in real world gameplay unless truly called due to appropriate fitting cause of challenge. This cause, as learned through experience, is prevalent in the actual surreal world of humanity. Games can permit Aspergic mindset insight into normalized strategies, useful as a means of identifying it better within the real-world during interactions with others, or as better recognition of injustice, harm, and manipulation in the motives of surrealistic persons differently contended.

An Aspergic person can be intensely skilled to perform and maneuver in such a world, but lacks the motive for the rewards, or prefers different rewards for inspiration elsewhere, even if not readily apparent, as an unsought challenge confidently trailed and trusted in uncertainty. In cases, it can manifest in recompense to those deserving with the prior confidence of will and current motivation to play such strategies in the world against its own species and all other species as an indifferently rationalized pride above the mainstream in the context of allegedly undeniable success of dominance.

–

Recognition can end opportunity for growth in defining ongoing beliefs and the emergent intuitiveness which surfaces from its wellspring of insights. The perceptions of others knowing a difference will shape the occurrences of identity which determine the Aspergic individually not much differently than any other becomes shaped by environmental influence, except in the scope of its intense presence between individuality and surreality as contradiction.

Often, Aspergic perceptions will find patterns within reality which the conscious mind will not become aware of until later under pertinent circumstances with a sense of becoming found while having unconsciously initiated somewhere before in the environment, usually with remembrance as a sense of familiar reoccurrence. The experience envisions, just as how a gene might appear benignly at one time, only to find essential use in regulating later gene patterns into a new solution otherwise unworkable without the earlier, previously inert gene being already present. A similar description could interpret for viewed fundamental particles of physics as an evolvement of solutions.

–

Aspergic tendencies could be a genetically regulated advancement; or an attempt at finding a path back from a normalized surreal world to a renewing association with a natural world; or a regressive deregulation for what becomes no longer required within a niche interpretation originally instilled by nature.

Aspergic talents recognize the spirituality in solutions of nature just as shamanism of ancient wanderers. Either is of equal value and difficulty whether perceived with a resolved, leaning preference towards real or surreal physical mysticism. For the betterment of the species in past environments, not all individuals are equally physical or mystical in their attentiveness except through the long-stringed attachment of forgotten origins.

Certainly, there are environments where the physical world reigns supreme over an absently perceived mysticism as an unchosen state, just as there are affordances for mystically bent individuals matched to sacredly insightful environments as dynamic enlightenment, as a type of idealized attunement of bodily survivability.

The spiritual experience can be of greater or lesser value than the physical one, yet over the range of its time-based totality usually assumes completed, poised equilibrium: even if not fully stable within any one of countless individuals observing endless moments of specific time and space.

—

A lack of surrealistic admittance of a spectrum of differing empathies is an Aspergic difficulty to contend; an obstructive self-admittance of the shiftiness of beliefs without counter awareness made through personal identifications of reality. It is difficult to admit to oneself what others cannot observe without the direct prejudice of prior knowledge; it is not as easy to guess at as it is to be suspicious of as an expected mode of surrealistic response to the nonexistent, unsuspecting unknown in all its vigor and telling scents.

Aspergic tendencies create a contrast which promotes conflicted responses of emotive discernment towards reality, increasing the probability of a revolting, contemplative, torn and dynamic personality mostly unobservable through normal, expected depictions of others except given the accusations against their actions.

Recognition that intuition can be fooled as easily as ideas can coerce a mind to act due to the inherent, malleable exertion required in balancing mutual understanding, words can be difficult to use as an implement for explaining ideas, since the interpretations available for other minds due the use of words can often create confusions of perspective according to divergent past intentions, experiences and pent resentments.

With Aspergic tendencies advanced within an individual, guessing is especially difficult yet still essential to social interactions. It is all guesswork against expected intentions as a predictability of ongoing experiences, and the dispiriting conflict of inconsistency between a singular identity and a disparaging, constricting mainstream.

As an avoidance of rudeness, an Aspergic personality does not initially or boldly offer to others what predicts in return as an invitation to brave rewards of exchange.

–

There is an urgently growing passion to recognize identity in the environment, but Aspergic influence is short in the world as revealed by aging readings of past and current human history where the Aspergic personality cannot find proper connection with its proportionally integrated nonappearance within human society.

Strangely, as a method of deflecting simple, subtle dishonesty, an adversely normalized surrealistic value, there is Aspergic comfort in being unobserved, in being without expressive opinions even as imitation of integration within plainly sighted multitudes of concentrated sentiments and interpretations, and the desire to further its sating lack of detection as an involvement with such whirlwinds of compulsory, socializing anarchy.

A nurtured reaction can arise from the obligation to not reveal the discomfiture of identification to others as an inevitable stone to carry, and as it is destined to recur in a normalized world minimally persuaded by Aspergic urgings for betterment.

–

Avoidance is a need for trust and certitude not found in the current environment, but in the environment of identity instead of sanctuary. Words, actions, and emotional arousals which express such certitude reach out with greater trust, prominent against a backdrop of empty actions, polished words, and the presumptions of self-aroused excitement.

Possessed of personas which are not so much viewable from the outside as they are experienced as a secret privilege is a splitting, uneasy state to dwell within, especially when contextualized within a reality normalized by contrary states of mindsets perilous in beliefs and actions, making of reality a seemingly more dangerous place to exist, far beyond the immediate risks of survival once bound to the natural world for sustenance, the attunement of skilled traits within tightened groups and inspired, instinctive individuality. For the Aspergic persona, surreality can appear to be a tightened constriction suffocating the natural integrity of instinctive values.

Distancing from others can be a sympathetic reaction symptomatic of Aspergic empathy, one of overall wonder at the emotions of existence primally shadowing the mutual affectations of humanity as an endeavoring, forgotten rapport.

Often, there is resistance to saying what should be obvious, acting upon what action should happen, or the calculation of what should occur as statement or action due to inadequate personal evidence and the pre-established predictions surfacing from haunting past experiences.

This resistance is too unreceptive and avoidable within the chaotic external world, forcing the inevitable result of needing to quietly watch what misses placement within surreality with its confusing motives of thought and intending beliefs of reality once easily fathomed as a naturally motivated predestination favoring humanity. Often what becomes overlooked happens with a profoundly aware purpose intensely safeguarded and usually unhindered by implications.

Individuals can be constrained to look further forward instead of simply backward in seeking a practiced anticipation of an eventual emotive destiny: through other means of strategy beyond the normalized, shortened immediacy of a human edifice as it exists to an Aspergic perspective as a context for needing a more genuine closeness of experience with reality, as an involved participant and attempt at coinciding experiences without the challenge of a human-instilled handicap. The Aspergic mindset must also contend with the delicate association of experience to resistance, and the loss of rewarding benefits should its peculiar hiddenness become diluted by exposure to the blending nature of a unifying, surreal acceptance.

Aspergic tendencies lean toward suppressed detection of the accurately divulged realities and integrities of identity, contending against the surrealistic credentials of humanity as a species in context with nature and reality in its incompleteness and brittle durability. Such tendencies conflict with humanity's surreal creation of itself as a cresting epitome of physical and spiritual dominance.

Such cognizance describes a depth of the viewable in deliberated experiences through the assuming aptitudes of one species within an otherwise impossible reality. Perception becomes communally shared to omit the burdened delay of deliberation, through calculating attempts at associative insights through the shiftability of the everchanging world, with attention for the consequences of the unfathomable as a realizable embryo, while possessed of an attentive, emotively introspective relationship with experience, not just humanity, nature or nonhumanity but the wholeness of reality as if existing long before the urgent need for the prerequisite of personalized identity.

There is Aspergic acknowledgement such poise should be naturally endowed and easily uncovered through a confident belief in personal action performed in attuned conjunction with a realized existence without alternative or the asking for more beyond simple betterment of survival, and a niche acceptance of equalized mind, body and environment bonded to the universal forces buttressing its successes.

–

Aspergic mindsets may tend to systematize reality's dependable associations, including emotive, social and identity realizations. It becomes easier to systemize the facets of reality than the duplicitous performances of humanity broadly prevalent and beyond true self-conceptions, which creates an anonymity in dealing with the world which disobeys confidence, and thereby reciprocally blunting or enhancing chosen actions. Sometimes this multidirectional maneuvering becomes lost to chance through the simple temporal expenditure of calculating the vigor and evidence vital in having conviction during personal choices, and the requirement of correctly deciding action and inaction without slothful obesity in justifying convictions.

As an unnatural presence within the crowd, Aspergic tendencies must assume the disguises of surreality without wanting to appear disguised. The natural world does not fit as a costume any longer to humanity in its current state of bodily social evolution, making vastly ranged populations strangers to the natural world of their own origin and birthright, and the many instincts and associations of passions gifted as an actuality for the confines of a species within a created, shapely environment, and within an individual of a species fluidly awed within themselves at the candor of a naturally occurred existence and an alleged fitful place within it.

Aspergic tendencies can appear as a natural construct for the potential of nature's self-adoration through the absorption of such a reflective presence in a perceptually blank, felt stare, like a mirror for reality's pride.

Aspergic tendencies prefer to become emotively attached to this mutual intention between reality and the experience of perceptions, as a bodily and mindful representation of this mutual incredibleness, as a living texture and coalescing exhibition of a recollected hiddenness of reality and the evoked power of silence. Minimally recognized among multitudes, Aspergic seeking finds agreement with this hidden silence as a shared memory of world success and a whispering empathy floating the waves of reality.

Sadly counter-intuitive, a proportionality of Aspergic persons is bound to become lost in the complexities of a surreal world and later defiantly averse to the hints of the natural world as an ill-fated outcome.

–

Imitating has a successful prevalence in nature and societal expectations of real and surreal normalcy, from disguises of body and mind, concealments of milieu, the estrangements of natural environments and redirected bodily stimuli into believable actions and responses, and as defenses against behavioral intentions due to the intrusiveness of the world beyond the individual by whatever shaped design. It is not solely a human phenomenon, though humankind has widely assumed its pride. It remains unowned as a familiarity except as a characteristic of satiating a niche outcome within a natural world.

Even in the absence of mirroring, there is recognition of imitation in others and recognition of its seeming absence in oneself. Also, there can be a detection in other people of a nonmirrored rejection of this absence or non-attempt at an expected simulated response: conflicting no different than its absence solely recognized as an absence without appreciated attention to the mimicking of purposeful normalized expectancy as a strange, surreal obligation.

This mirroring, conceived as a trait of survival, has its own budding advantages stemming from its accepted recognition as a trait through the intentioned compulsions of benefit within the imposing compositions of everyday life.

With Aspergic inclinations, an individual is self-observed not mirroring alongside others through the subtle, postured reactions of comforted reply, but as witness within a personal, Aspergic context of the absence of the trait not observed in reply as expectations within surrealistic minds driven by the reassuring predictability of configured, genetically regulated social echoing.

While perceiving Aspergic tendencies, surrealistic perspectives can be conflicted as they try to naturally imitate a trait which is non-expressed or differently expressed within the normalized individuals due the confounded absence of pleasing rewards motivating properly satiated anticipations or, more appropriately, a red flag to its affronted expectations within an immediate environment epitomized by a challenging, unfixed Aspergic presence.

These conformities to convicted expectations in social behaviors are the appeasement of inconsistent integrities resulting from the individual's capacity to predict the repetitive world with differing experiential capacities, understandings and skills in ongoing optimism of correct responses from a preferably nonconflicted but often conflicted world in its relationship to the experiences of a defined personality and any attempt at embracing a retained identity.

Aspergic tendencies can be an affront to these reward-seeking expectations and the comforting grounding it supplies as mutual reexperience of pleasing reassurances within a social playground. Obviously, rarer individuals not perceptually comforted in equal measure by the same rewards of social anticipation through imitation, seeking and finding alternate means of self-approval with the world, will become exceptionally observed and naturally suspect of questionable, even disloyal rebelliousness as an opinion of an appeasement-seeking, surrealistic attention.

These attempts at mirroring can seem more a struggling type of mocking rather than as imitation from the Aspergic-inclined perspective, as a feature to learn to acceptingly understand even at the exclusion of oneself from the equation if that is the highest probability of overall success in retaining a sought neutrality.

No different than the mimicking of the mirror impulse achieved in mutuality as reciprocal recompence is prevalent in a normalized social world, it will find resistance as an attempt at exchange in an Aspergic-minded personality, forcing an unexpected disagreement with surreal normalcy against such a determination to observably outwill it in its observed act of contempt to decorum; rather than continue accepting and acting upon the original, disagreeing impulse not found profoundly existing other than rigidly within the habituated expectancy of surrealistic minds and their alleged preordained world of erected, idolized creation perfectly customized for proven human mechanization.

A surrealistic mind trying to mimic an Aspergic mind through accidental environmental interactions can have difficulty with its legibility. This recognition can be bluntly affronting and re-represented in return creating unnatural reactions enabling the higher prospects for conflicts resulting from the surrealistic mindset's inability to recognize the Aspergic mind as something contrarily distinct, just as nonhumanity construes as indistinct in its highly unifying awareness absent through the latency of nature's undertone.

The Aspergic presence must learn to contend with an overwhelming external difference while the surreal-inclined mindsets can often live entire lifetimes without a desire to experience or believe in this difference allegedly fathomed by others as if a mystical disbelief in reflective physicality like an illusory manipulation of light.

This relationship can be a potentially profitable pursuit of associative learning and sympathetic understanding consequential to the necessity of utilizing perceptual Aspergic skills over the same lifetime as a stimulating, unceasing resolve fixed within an otherwise conflicting reality in need of successful contention as a normalcy of survival.

This necessitated analysis of the environment for cues and information can be an added benefit for a proficient, Aspergic mind in need of better connotations in understanding contrary normalized, surrealistic tendencies and beliefs of reality as an overwhelming inequality.

Each identity becomes mutually unchained in spirited bias of sensual withdrawal, in their confused inhibitions in reading another identity and the believability of opinions over observations. As an opportunity of unification, the coexistent Aspergic and surrealistic natures can also mutually empathize, even if this empathy solely encounters through amiable understandings among a minority of individuals: gained through the shared associations rewarded by a close commonness of afforded experiences unified through a coincidentally bonding spiritual, emotive, or believable motivation.

Aspergic pride inevitably needs to win out at whatever costs or resulting gains crave to occur as reparation. Aspergic tendencies force an individual to try actions meant to socialize and to do unnaturally what a vast mainstream of surrealistic dispositions finds natural, with reassuring consent and adept skills required of its vibrant social arena exhausting the dominating surreal world.

This breach of opposing ownerships enhances the difficulty of self-enabled Aspergic gauging of predictable scenarios of actions expected of its world, as if a non-actress among actors, often misconstrued through social normalizations as timidity, introversion, or vague frailty of character spooked by an obvious surrealistic detection of internalization cast from the Aspergic persona.

Through a lacking perceptive aptitude within many surrealistic sensitivities while prevalent within Aspergic aspirations of reality as a different, unhindered normalcy, there is a mutual mistrust of intention due to differing perceptions not needing to find acceptance within the surrealistic mind as readily as it does within the Aspergic mind as compromise according to the natural imbalances of encouragement embodied within the normalized structures of surreality.

The disproportion is the evolutionary skirmish of biased corroboration within the world's designs. It is not just a lack of understanding but a lack of acceptance of what can be understood, and a lack of willingness to burden with the requested empathy for the deep implications of tolerating manifold versions of truths, compassions and sighted beliefs as a seeming unity of its spiritual perfection within the physical, environmental world of surreality in all its reoccurring outcomes.

This is meant to become circumvented as an unsuccessful path to best avoid for the betterment not of humanity or its individuals but for the instillation of forces capable of furthering the capacities of the species persuaded by surreal encouragement of reality to bitter ends.

The disharmony emerges within the purposes of direct and indirect communications and bodily relationships of posturing and counter-posturing, forcing an Aspergic persona to learn to be consciously secretive of unconscious impulses happening within the environment, enchantments of what others really do not want to hear or observe as a likeness of truth contained within a solitary, staring observant presence withstanding crowded, sensory disquiet with a steadfast resolve not to cede a colored petal of self-possession.

–

An Aspergic person distinguishes between sincere and disingenuous in the flowing interpolations of experiences. This recognition can collect observations within the most pertinent memories retained through an advancing lifetime for reflective prophecies of expectation, such as should be predictable to surrealistic minds from the Aspergic perspective; instead, repeatedly observing an apparent lacking or absence of sightedness for such finely lighted spectacles.

Natural symbols of instruction remain for future use from past practices, and the Aspergic identity inevitably learns to apply likelihoods as a compensating solution to the necessity of social interactions, reckoning with performances, motives, intentions, and risks as a persistent ritual of enabled surrealistic attention.

The Aspergic person can come to overlook themselves in this perceptive game of educated guessing, whereby the world can seem to exist as something *including* them in relevance identical to a freeing, pressing sense of personal exclusion from it as an identified, fragmenting right of spatial independence.

–

Aspergic tendencies try to act on motivations not exactly for personal gain or redemption but for overall betterment beyond oneself as a whispered echo and tendrilled influence on future outcomes. Sometimes, this betterment of reality is not yet plain but merely a disenchanting envisioning or prophecy not readily sought or yet presented within a strangely surreal reality demanding reckless self-attention of its individuals given to one convergently multi-devised interpretation.

This intended betterment can also be illusive and is prone to premeditated, divergently ill-fated penalties surfaced from the past as an endless, nourishing consummation within the present state of the world.

–

There are private aftermaths embedded within individuals according to the benefited extent of livingly robust, visual memories as a possession of mind continually seeing the colored details of a tidally conflicting surreality placed atop reality. These memories endure as an obligation of survival for the Aspergic person incapable of merely generous recollection. Memories survive according to the value to the one remembering, so that they become important or not based on the same implements of detection or imitation of the past in context to the memory's sole ownership within existence as a potential avenue for advantage by exploitation. Memories become a means to an analysis of a personal history and its merits or basis of contention as an identity. Memories can also become altered to reflect a false sense of personal history and empowerment within a caricatured, unproven, or victimized individuality.

It is rare for these distinctions to unveil within the fabrics of surreality expressed by its personas; instead through the concentration produced by inspiring books, art, the looks of nature and the collected, reflective insights of experience within reality retained through discovered, inclusive events as living aspects of the past.

History as probing feedback can become a fitting source of discovery, learning the exuded proclamations of similar confined, improperly fitted Aspergic presences as they coped and summated the reality's value. The same world they experience during any of countless strange epochs of time, with the vigor of lifelong rebellious personas trying to find a face of identity within a spiritually misrepresented, shadowed atmosphere of conned worldly identity.

Aspergic minds can look to the past for source of solidarity, individual exertions of will and striving, and for enduring minds lived in unknowing contrast and without proper context within an ordinary, temporal world.

–

There can exist a sense of solitary disassociation with the world of moveable occurrences around an Aspergic perspective seeing not only the results of behaviors in the world but the impulses, motives, designs, and insights of the alleged processes causing them. They feel and sense the inner workings of natural reality as betterment without the confidence of certitude in oneself as a fitting source due to minimalized encouragement from a preferably suspicious, contrarily perceptive world differently explained. For its full appreciation, this intuition needs attention throughout a lifetime of experience and self-reflection against and alongside the unseeing contrast of inversely disposed multitudes.

Due to the negligible inspiration of Aspergic tendencies within surreality, the overwhelming cause of the disparity felt and sensed is a neurotically infused world of overconfident actions and unrepaired consequences without truly genuine foresight, attention, or predictability observable as inner workings of its world. Instead, it exists as a world prevailing without excess sensitivity beyond or destined to the moment. As a context, it is visibly a disobedience to a previously teeming natural world and aged universe before the advent and tussle of personalities surfaced as a minor, misled contention of true proportional value, especially without proper measurement scaling a member of a species against an infinitely separated cosmos.

The world has amassed a multitude of intricate details with unsorted connections without any conceivable or realistic overall bigger picture. The big picture of humanity is one of perpetual conflict over rambled fantasies inside its secured, embracing surreality, a world constructed as guard against the overwhelming exposure of the universe upon all chosen, isolated actions in belief of its supremacy over a natural planet brimming with aged successes, against a natural world of private nonhumanity, against its own species and generational futures of offspring, and as reply to a cosmos quieted to its secret, permissive and satiating pleasures.

Focus and rightfully perceiving specific interests and observances of reality can act as a method for clearing the obscurity of many small details into something tangible and real, much like physical matter clings to the preestablished forces of an endless summated past for support and substantiality within the same universe, shortening its focus and permitting evolution as the spirited vigor of actuality in all its gained poise; beyond a shrunken consciousness limited in its ability to observe the world further than its abbreviated facts and definitions masking a self-reflecting schema of control.

—

A residing spirit freed of the human-erected world to the extent of a journeying self-deniability contrary to its confidence, as a lively source of wonderment or insight for the identity of the individual, as something other than renunciation of the natural world in favor of imaginings of cosmic spiritual favoritism, is the tendencies of the Aspergic-minded person in the pursuit of interpretations as a surfacing relationship with existence.

Aspergic tendencies must learn to deny presumed attention and opinions of reality to enable private social influence upon emotive thinking to be the mere appeasing of mainstream expectancy. There can remain a lingering sense of doubt at the effectualness of these internalized denials as a self-corruption of skills at having to simply play along. It is most difficult when the Aspergic person finds no cooperative support for brands of interpretation different than that within prevailing local and broader temporal environments of opinion.

In some cases, deniability of surreality can become widespread within the cognizance of an Aspergic person, conflictedly learned as a confident self-fulfillment, the capacity to overlook oneself as an acting aspect of the coercive influences of a species, society and an environment lacking in true empathic explanations of the world, as an urgent, unappealing and unremitting craving for an enacted separateness from ownership.

An Aspergic personality can be coerced by the prevalent conditions of a world to spend an entirety of a life taking on its inner makings, enduringly rebellious, thoughtful, and silently inspired to transcend the contrariness of the surrounding reality perceived as textured differently than it is presented and believed through intimate experiences with normalized social adaptations.

Such minds must learn to understand when and when not to act on what can be at times an overvalued reading of individuated firsthand intentions and emotional states when merely occurring as an internalized awareness separate from the event, creating an individual who must control emotional impulses and mindful states internally envisioned without true ownership of all that the surreal tendency incessantly represents as its objective manifestations of pursuit.

This creates a sensitively complex, withheld condition within an individual in relationship with the world, as empathic observations can lead to the sense of needing detachment from self-involvement due a neglected sympathy to what is beyond controlling, avoiding personal connections as beyond the scope of being an influential identity within the varied inactions, motives and designs of the human-made world, thereby adding another deep layer of separation within the Aspergic person through self-designed relinquishment of ownership of the world, as if the world really did not belong equally to all in its representations of admittance.

The individual must extract these integrities solely through the experience itself as an acute alternative that persists as discipline against the receiving of events otherwise deficient in mutual attainment.

Aspergic minds become an overcompensating origination of an extreme and risky environment beyond any conception of nature in relative scope. Without this condition of environmental influence assuming such a dominant comprehension of an otherwise unimaginable world, an Aspergic mind, no different than the surrealistic mind, should evolve as a force within reality along separate and differing waves of represented, composed revelations.

Though, as recompense, with the current world existing in a current and past state of misinterpretation, the seeking Aspergic identity finds as rewards its utmost talents, whereby reality itself becomes the artful canvas for the deft strokes and storming colors of emotive, mastered artists bursting with secret sightedness.

–

It can at least be imagined the many individuals throughout human history until the most current present state with varied degrees of Aspergic tendencies yet living without knowledge of a difference or even the notion of a differentiation existing as a possibility to be considered within human sensitivities and its eccentric new world of acceptable ideals: and the temporal fears associated with such bold revelations among ill-informed listeners. Of these individuals, there would be the few to succeed and the others to fail against the struggling pretense of a social personality contrary to an intuited nature: a subjected identity inside an objectified surreality while beyond a natural, justifiable context of support.

They have lived in a world while perceiving existence as surfaced differently and consistently without thinking it anything other than normal in the staring eyes of an inquisitive persona colored within the usual inattention of everything else in widespread black and white, surrendering to calls of victimizations, mental afflictions, tortures of forced concession as compromising penalty for any possessively raised awareness of social confidence.

In such a mindset, any idea of contradiction is overshadowed by the varying ideas of the spectrum of equality and misrepresentations of reality due the relative inexperience of human history determining its actions contextualized against the ranging scope of current experience, a present and past state melded with its exhibiting penalties and apathetic devotions not much further advanced in fathoming individualized distinction beyond rewarding humankind for its purported skilled victories over a planet and its common nature.

This event occurs in the present state of a reality, during an era of abundant data, science and claims on freedoms and awareness due to the progressive evolution of humanity's unwitting accumulation of disproportionate latent knowledges it contagiously extracted as origin and erosion of its environment.

There should have been shamanic mystics, originators, leaders, sceptics, and artists, unsolicited by the myriads of potential niche atmospheres they would have been unwittingly born into as the means to an end for rudely acceptable surrealistic ingenuities of imagination, beliefs of unkind insights, and plentiful self-appeasements.

Also, they would have received punishment, segregation, shame, and treatment as discipline through the commands of a normalized mode of thinking, confining its undying depiction through the disassociated approaches and services executing a surrealistic-dominated human history with its slew of famous perpetrators.

These forceful old costumes of history persistently remain, directing human choices in the present state of the world, as diagnosed through the focusses of a moveable Aspergic lens upon reality.

–

Due to ineptitudes within the spectrum of knowledge instituted across unfamiliar generations of history, attitudes of the past cultivate an instinctive social dichotomy within those Aspergic persons undiagnosed and misrepresented, existing in countless historical milieus, during recent and current times as an enigmatic, metamorphic wanderer and rebellious influencer upon budding embryonic symptoms within the spectrum of reality as it is associated with human social history, hardly counted among countless localized reoccurrences nudging its evolutionary attentions in jointly new and lost directions.

Unless a personality is fully consumed in repressive imitation and disguise accorded the stringent principles of social enactivism, an Aspergic awareness observes the experiences and the makings of reality's symmetries as if mystical, or the convergent impression of beauty as if an observable ingredient of lighted, surfacing and nonphysical exhibitions emanating as the guarded modesty of nature; or the summation of Aspergic identity enacted through the seeking and retention of lively patterned associations inclusive of nonphysical manifestations naturally cultivated from reality.

An Aspergic individual does not naturally see peoples or nations, except as humanity and reality. The Aspergic tendency is a willful attempt to find order in humanity's emotively complicated disorder, and the world's social and intellectual means as a frustrated, driving need for communal solutions beyond the indulgences of the ego-self as an individuality within highly populated groupings, rather than as an internalized illustration of unbiased empathy due the curtailment of Aspergic identity as if desperate, rebellious avoidance is an only ready solution.

This attempt at escape finds its redemption as a losing surrender of one attached reality in favor of another attached reality reciprocally isolated in context whether in a rallying human arena or a secluded natural space. In either case, the sense of detachment is the same, in whichever experienced alternative of action becomes choice, so that any idea of escape entails nothing more excessive than a change in perceiving surroundings as an influence of direction and revolution of desires.

Nature is emotively releasing for its innovative sincerity upon Aspergic susceptibilities, as if by identification mutually recognized in common, and as a separation given a human-instilled withdrawal from the past and future as relevance to an occurring present world. Nature is proof of truths not found or else lost within the human, surreal, idealist spirituality, in favor of other purer symbols and actions rooted in embedded, private, or esoteric beliefs and ravings.

It is difficult for a normal thinking mind to understand why an Aspergic person would not want to adapt to the surrealistic-favored world with its multitude of acceptable and contented benefits.

It is difficult for a normally emotive, bodily astute mind to fathom how an Aspergic person should remain as unfitted for the world it rebuffs to the point of revolt across the spectrum of a lifetime, as an accurately fitting, chosen response while arrogantly rejecting the common, welcoming gesticulations of surreal opinions.

If everyone else behaves with solemn believability according to conformed ideals, then it can be confounding why an Aspergic individual would not want to become closer to those communicative influences through acting like or preferably learning to appropriately imitate such a world it finds inherently difficult to appreciate, and a condition of personality repressible for reciprocal betterment to individual and the greater surrealistic majority.

As with evolution, the societal world desperately seeks better solutions to further its evolutions, with an equated range of deniability and comprehensions much as with species facing changing environments, whether composed of an original, bodily, or mindfully surrealistic world.

The prism of Aspergic beliefs views a world both real and surreal, riding the stormy waves between the two realms of manifested truths.

Aspergic sensitivities readily sight the natural world with capacity for reaching far greater depth than the incongruity of a distracted humanity due to a proficiency to focus, to single out and associate, like predator to prey, momentary experiences of realness within surprising, unpredictable, surging instances of merging contexts, through the valuations of unmasked evidence and senses of reality found plainly hiding as rare but discoverable, lucky gifts.

–

Over the range of a lifetime, an Aspergic individual can consume inordinate amounts of energy stubbornly trying to rightfully prove oneself valid against a backdrop of self-doubting sureness instilled by an inversely composed world forced into unsolicited contention by the forces of universal evolution.

In a world ripe with fruitful revelations, the Aspergic attention can come to limit what it reveals to the world due the recognition of risk at what cannot be sufficiently gauged by a disinterested, surrealistic-persuaded world, yet be completely haunted of a diverse empathy, more inclusive and less outwardly animated than other divergent, bodily endowed, intimidating personas found through the temptations of reality continuing the reoccurring compulsion of past generational abundances of successes through unconsented aggressions.

It is difficult to find true empathy in surrealistic-determined worldly actions and acceptability, no different than it can be exceedingly unimpressive when encountering the reactions of an Aspergic mindset enabled with such pure sensitivities but instead outwardly imitating the surfaced surreality according to its enhanced powers of regulated culturing, thereby merely impersonating the rest of the world in its replication of surrealistic-determined compassions.

Such minds can be vulnerable to losing a contextual identity of private empathy due to disenchantment of the external, normalized reality, making it contend with the world as if burdened with seeking consent for such abstract, genetically-enhanced insensitivity rarely otherwise found to experience except through the motives of personal involvements mutely exposed to the enabling aftermaths of surreality in its countless past and current enactments, churning its justifications founded on an original ignorance, furthering its rallying opportunities for advantage.

The Aspergic individual becomes compelled to explore the thresholds and boundaries of reality, to push outcomes, stirring up new descriptions in its pursuit of assessing the murkiness of its environment for essential feedback.

The Aspergic individual commands the choice to assume and become consumed by normalcy, or carefully embrace the non-normal sensitivities afforded by a separate way of thinking, feeling, and sensing the world. There is recognition these differing sensitivities, the same as with nonhuman species in their originally landscaped world, can share the manifold points of view as willful appreciation of the spectacle of varied awareness vital for an overwhelming, far-reaching success of global survival.

A diversity of experiences is essential to a relationship showing associations of patterned thinking gotten through a striven, emotively advanced learning requiring the detection, avoidance, and corrections of individual errors as an accumulated, kept new attention to reality. Aspergic tendencies inspire individuals to learn to tolerantly act along with regularized insights and tempered ideas as the methods of surrealistic-inspired survival, even if it only applies superficially and with applied caution as true inspiration from the quietude of an Aspergic perspective.

Within these individuals, normalized and victimized by the inborn chance accident of a surrealistic-designed reality, the inner world can be an unaware conflict of differing perspectives, patterns of understanding and depth of associations versus what is normalized as *acceptance* in others when it comes to understanding justification for behaviors and the customary, worn methods of ready opinions of reality. It can become difficult to reconcile what is intuitively perceived distinctively within the natural world with the surreal edifice of humanity attempting to raise itself above it from its starting ill-informed, absent awareness founded on instinctive cognizance and a varied, spittle environment of pristine, contending symbolic aspirations provisioned for its newfound consciousness to become seeking and playful.

–

Among individuals the differing awareness can blend and appear seamless when the influence of an external supplementary world allows a normalization of Aspergic personalities to glow and freely, persuasively texture a reality.

Aspergic tendencies are one small aspect of the spectrum of personality, or the spectrum of reality's coalescing forces, or the spectrum of awareness among nonhuman species, and the measured spectrum of dominance within environments. Among all these coalescing spectrums of a world, the surrealistic interpretation contends as physically and spiritually absolute, and as best fitted for challenging this willing, defenselessly consensual natural world, with its corresponding universal forces harmoniously scaled as equivalent to the energy of infinity versus a planetary speck of existence. This scale corresponds to the dense, speck-like stimulation of Aspergic persuasions enacting change upon the loud momentum of an assaulting, surrealistic-concepted world-ordered history.

The Spectrum of Personality

The spectrum of personality allows the accumulation of a range of permissible talents, skilled awareness, diverse acceptances, and intended beliefs. With Aspergic tendencies pulsing within an individual's desire for better identity, a force must become exerted and accepted as received before it can be of meaning, thereby creating a response in return. In a way, social energy can only disperse if fittingly absorbed from outside beforehand. Initiatory action is limited as viable alternatives for Aspergic individuals, reflecting best as a social success when balanced between identity and environment, usually occurring as localized comforts between intervals of discomforts with others less uniform in reciprocal responses – or contrary, unempathetic personalities – and the coincidence of being occasionally, sedately removed from solitude through another person, people, nature or environment as a reprieve.

The surrealistic personalities include within their divorce from natural reality a wide range of variables and potentials in their quest for successes, so long as a conscience is cleanly appeased through a provision of accountable networks of beliefs including the basest self-rightfulness. The question can arise whether this absence of variety qualifies as the best fitness within a socially believing and acting species among numberless species sharing a closed environment atop a bubble of reality.

The Aspergic individual can come to observe the motivated consequences of history as having created a self-fulfilling scenario, especially growing over recent millennia, into a reality where the surrealistic personality determines the future course for all of nature and humanity due to its prior predatory successes of physical feats of dominance, and its pleasingly rewarding detachment from integrity versus alternate, contrary originalities of an overabundant world ready to be willfully sown.

Individuals of conflicting cultures will choose to forcibly adopt abrasive, dominating, and narcissistic personalities, with primal behaviorisms as believably acceptable, promoting the standardizing of these personalities far more prevalently damaging to the course of human history as determined, disintegrated motivations for enlightened attainment.

Aspergic individuals can naturally question such motives within the world as readily unfitting, as demanding unattainable forgiveness, but often seem to act alone in this quest for insight.

–

Aspergic tendencies can arise within a small part of an advanced conscious species, and a small part of those learn to express and reflect it with any pronounced skill. The reason is genuinely a lack of environmental knowledge, opportunity, or interest, and the unnerving tendencies exuded by surrounding, excitable types of personalities, equally constrained by their own dilemma of mind, and often forceful in opposition to relinquishing an original, misconstrued ignorance of humanity as a fitting mystical origin.

For innumerable generations, Aspergic tendencies walked unnoticed, unrecognized, given to a normalized, surrealistic deniability in recognition of difference among regulated traits of personalities. Humanity did not have pronounced interest in personalities except as forceful appeasement to an associated reflection of iconic status within its natural milieu; and as fodder for new arising personalities to reassume. Accordingly, the characteristics of advantage in these tendencies have become accepted and integrated according to the urgency of its qualifications within goading surrealistic aspirations, predetermined tolerances permittable to surrealistic successes, thereby revealing various genuine impulses of surrealistic self-bias in its wake.

In this context, Aspergic tendencies show resistance as an interpretation challenging the true idealized beliefs of reality contained within the believable imaginings of its species. Without a sense of overwhelming justness, the Aspergic identity has difficulty pursuing the mainstream as an ideal representation of the world, instead of as a mired world counter-resisted in agreement.

—

Living among opposing conditions of personality, the Aspergic individual must learn and practice facing a given reality without private segregation, and thereby without loss of experiences against other differing groups or conditions of personality. This is a reciprocal gain since the surrealistic mindset must learn to cope with differing interpreted sensitivities of reality other than its unchecked original pride. Otherwise, there is mutual inexperience especially during younger ages leading to the same repeating behaviors in adulthood without reciprocal gain in understanding other than through conflicted resistance due an equalized mistrust.

Aspergic tendencies and other minority personas throughout history will have contributed to the benefit of the surrealistic world's dominant stance. Though difficult to quantify the contribution, the absence of these scattered insights, ingenuities, alternative patterns of perceptual talents should have limited and impressive consequences on the shape of the current state of such ready-made world, to its unwitting culpability as contributor to an unbound surrealistic reality unconstrained by genetic integrity.

A growing awareness of Aspergic realization expressed through its tendencies creates a moral dilemma concerning the difficulty with passing on genes. Observing a plainly viewable world of surrounding experiences and the consequences to personal identities and social dynamics in the immense trivial details holding totality, it can seem unconscionable to further a genetic journey. It becomes difficult to ascertain which is a worse world for an offspring, surrealistic or Aspergic in context, and whether the Aspergic individual can decide or make such a punishing choice for a future offspring to have to imprison within such an illusory, hardened reality.

Aspergic individuals look for truths, the right and wrong of an event or situation such as humanity on a planet, its implications, impressions, and deniability, not wanting to act on falseness or pretense as a contributor to furthering an existing misdirection.

For reasons of identity and the integrity of having to difficultly live with oneself and one's consequences with heightened recall and conscience, as the workings of an inner self-making, Aspergic persons cannot as easily become corrupted in status or personality. In this way, they can be inspiring for their impressive persona of an exaggerated kind of normality, though this exaggeration only exists due to the contrasting exaggerations of normalcy favored by surrealistic identities.

Perspectives of Aspergic experience as an aspect of human history apply to all contrary perspectives of feelings and beliefs among humans and nonhumans diluted in relevance against the dominance of a rallying surrealistic viewpoint.

There are certain to be varying marks and stimuli throughout history of such Aspergic tendencies; defined as a condition, such a thing would not just abruptly stop or occur in a species with such an absorbing capacity of imitating personality traits coaxingly enhanced by the persuasions of environments.

–

Surrealistic identities can become limited in focused attentions, intentions, and skills the same as Aspergic minds and bodies tend towards multiple interests of useful skills and practiced levels of skills depending on the motivations enabling a repeating willingness of pursuit. It is the hunt or success of seeking which is limiting due to success externally conceived more as a failure compared to an overcompensated expansion of surrealistic attention as priority. This can create an underachieving individual as a means of fitting in or dumbing down as a relationship with the world it needs for sustenance, the same as any surrealistic-tempered person.

–

At any point in the history of advancing an informed cognizance founded on empirical understandings of minds and emotional responses to complexities of being, with individual responses to a reality's replies to its societal stages of understandings and coinciding misunderstandings with each new stage of betterment, not just of the abstract obscurity of self in the face of an agonizing awareness of existence and the frustrated limits of grasping this seemingly unfathomable hiddenness, it can be seemingly normal to avoid this vastness in favor of a subdued or extravagantly misled, self-imaged personality.

Through the sluggish evolution of history's understanding of knowledge, these resentful conflicts have become absorbed by a storied humanity and nonhumanity, multiply exposed to a hard reality for such minds to fully grasp and accept, yet to sense with high awareness something lost or never found due to an off-centered loss of imagination.

Normalcy in This World

The natural world promotes diversity among species within environments and among individuals within species. This variation allows limited behavioral freedoms through a genetic exhibition within reality promoting subtle differences according to physical capabilities and local variations of environment. As a consciousness grows within a species as a deterrent for better survival, there are extended residual emergences from these communicative idiosyncrasies: as a consciousness begins not to shape niches but construct them as a diversity of appearance, taking on in struggle and resistance the entropic reality it initiates its new surreality upon as a backdrop of symbol, memory, convenience of use, and the underwhelming condition of its resistance to overwhelming will and force.

The natural world, learned through adaptation, can become readily outwitted by a properly evolved bodily shape with a coinciding clever, predicting mind. Within an unconscionable species with conscious will and bodily capability to shape reality to its bidding, nature is vulnerable. This inscrutable will exists within a bubble of disorganized and incomplete ideas without vast overall aptitude for understanding future outcomes as a forceful, mainstream independence of hungered obesity as its surviving ambitious desire for overindulgences.

To observe the real and surreal worlds in contrast and with contesting juxtaposition, a world where true integrities are disregarded in favor of opined views of reality, where differing perspectives are irrelevantly treated and acted upon through a resentful commonness of individual personalities untrained for the natural world of existence to foster higher measures of command and confidence standing before successful innovated, celestial landscapes, as witness to the natured personas of an open-minded universe segregated as a species from favoring a practical mindfulness empty of true mystical, coincidental and natural connotations.

Normal is not erasing the present from the future as experience. It is not omitting this experience from unnamed others, from upcoming generations. It is not removing the temptations for awe from the world. It is not the end of belief within the imagination of humanity as a spiritual and physical wholeness in isolation, rather than the opportunity at diverse enlightenment it ceaselessly afforded overlooking.

www.ingramcontent.com/pod-product-compliance
Lightning Source LLC
Chambersburg PA
CBHW060458030426
42337CB00015B/1639